A GENTLE INTRODUCTION TO GRAPH THEORY

notionpress.com

A GENTLE INTRODUCTION TO GRAPH THEORY

VALSAMMA K. M

notionpress.com

Notion Press

5 Muthu Kalathy Street, Triplicane,
Chennai - 600 005, India

First Published by Notion Press 2013

ISBN: 978-93-83185-63-4

PREFACE

The aim of the book is to introduce undergraduates (and perhaps higher secondary students as well) to the Mathematical area called ***graph theory*** that came into existence during the second half of 18th century. An attempt has been made to cover elementary to advanced concepts in each chapter and to take care of the needs of students endowed with little or no prior knowledge of the subject. The book is also appropriate for self-study. Each chapter contains sufficient number of illustrations with examples to explain definition, principles, and descriptive materials including theorems. Graph theory is an area of discrete mathematics that concerns the study of mathematical concepts and their inter relations. What makes graph theory interesting is that, it can be used to model situations. In graph theory, powerful concepts can be defined and introduced because they can be visualized and simple examples can be constructed easily that make the study of the subject more rewarding to the teacher and student alike.

The book is designed to be self-contained and consists of 8 chapters. It is useful for students of Mathematics, B. Tech, M Sc and MCA syllabus of various universities. While writing this book the author had befitted immensely by referring to several books and publications. I express my gratitude to all such authors, publishers, many of them find a place in the references. I am sorry if any such source had been left out inadvertently; I seek their pardon.

All efforts have made to make this text both pedagogically sound and error free. However I retain the responsibility of any kind of errors in the book. Suggestions to improve contents of this book are always welcome and will be appreciated and acknowledged.

Valsamma K M

CONTENTS

1

INTRODUCTION

Many concrete practical problems can be simplified and solved by looking at them from different points of view. In the recent years, there has been significant change in the relationships of mathematics and computer science. Earlier mathematicians helped in designing computers for the purpose of simplifying, their own complex computations. But now the more specific needs of computer scientists are evolving a new way of doing mathematics. Graph theory or study of graphs is done by computer scientists because of its many applications to computing, data presentation and network design. Our journey into graph theory starts with a puzzle that was solved over 250 years ago by Leonhard Euler (1707-1783). The so called Konigsberg bridge problem, was a long standing problem until it was imaginatively solved in 1736 by Euler. Konigsberg was the capital of East Prussia. The Pregel river flowed through the town of Konigsberg. Two bigger islands protruded from the river. On either side of the main land, two bridges joined the same side of the main land with the other island. A bridge connected the two island. In total, seven bridges connected the two islands with both sides of the main land.(Figure 1.1). A popular exercise (today's logistic problem) among the citizens of Konigsberg was determining if it was possible to cross each bridge exactly once during a single walks.

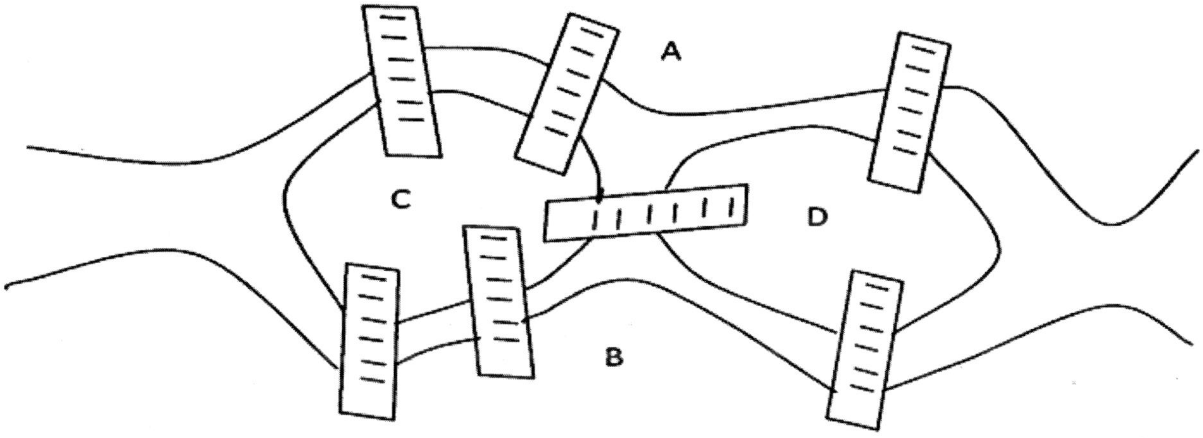

Figure 1.1 The bridge of Konigsberg.

Euler's was to realize that the physical layout of the land, water, & bridges could be modeled by the graph shown in Figure 1.2.The land masses being represented by small circles (vertices) and the bridges by lines (or edges) which can be curved or straight. By means of this graph the physical problem is transformed into this mathematical one. Given the graph in Figure 1.2, Is it possible to choose a vertex, then to proceed along the edge one after the other and return to the chosen vertex , covering every edge exactly once? Euler was able to show that this was not possible. Euler solved this problem in 1735 and with his solution he laid the foundation of what is now known as Graph theory. In graph theory one uses mathematical structures, to model pair wise relations between objects from a collection, that are related to each other and these structures (graphs) are used to model a lot of real life problems. Graph theory is now an established modeling method used in a variety of disciplines like Ecology, Geography, Information Technology, and Computer Science, to describe relationship between objects. In this introductory chapter, first we provide an intuitive background to the material that we present more formally in other chapters. We will also discuss some of the basic results and theorems in graph theory.

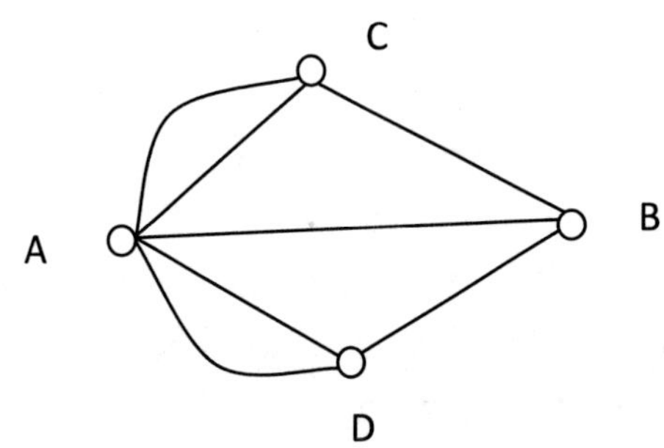

Figure 1.2 A graphical representation of Konigsberg bridge problem

1.1 WHAT IS A GRAPH

Before we can begin to deal serious concepts and theorems in Graph theory, it would be interesting to find out what really is a Graph, how it comes into existence and how does it relates with other areas in science like, physical, chemical, biological, social and numerous other areas like, linguistics and computer science . In this chapter we briefly outline these issues.

We will define a graph as an abstract mathematical system. In order to provide some motivation for the terminology used and also to develop, we shall present graphs diagrammatically. Any such diagram will also be called as graph. i.e., A graph is a drawing or a diagram consisting of a collection of vertices (interconnected nodes) together with edges, joining certain pairs of these vertices.

Having used the term graph quite a bit already, it is time now to define the word properly. We start by calling a "graph"what some calls as "un weighted, undirected graph with multiple edges".

It is a fact that many branches of Mathematics begin with sets and relations. Indeed, graph theory is no exception. It studies relation between elements. Mathematically, we can write,

A graph G is an ordered tuple, G = [V(G), E(G), φ] Where V(G) and E(G) are two finite sets defined as

V(G) = Vertex set of Graph G.

E(G) = Edge set of graph G such that each element e of E(G) is assigned an Un ordered pair of vertices (u, v) called end vertices of e.

and

φ = A mapping from the set of edges E to a set of ordered or unordered pairs of elements of V.

We denote the graph G as G(V, E) or simply as G. A graph in this context refers to a non empty set of vertices and a collection of edges that connects pairs of vertices. The set of vertices is usually denoted by V(G) and the set of edges by E(G). The most common representation of a graph is by means of a diagram (as we did in Figure 1.2), in which the vertices are represented as points and each, edges as a line segment joining its end vertices. This diagram itself is referred to as the **graph.**

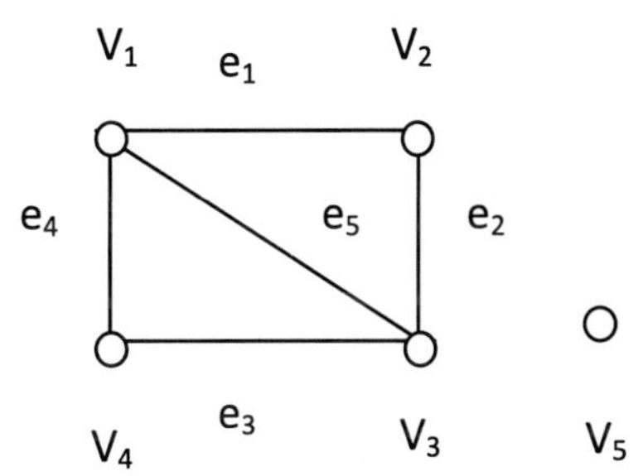

Figure 1.3 Graph with five vertices and five edges.

Thus for the graph of Figure 1.3, the vertex set is V(G) = $\{v_1, v_2, v_3, v_4, v_5\}$, edge set E(G) = $\{e_1, e_2, e_3, e_4, e_5\}$ and φ is defined by $\varphi(e_1) = \{v_1, v_2\}$, $\varphi(e_2) = \{v_2, v_3\}$, $\varphi(e_3) = \{v_3, v_4\}$, $\varphi(e_4) = \{v_4, v_1\}$, $\varphi(e_5) = \{v_1, v_3\}$. Another typical graph might be a family tree where vertices are persons and an edge connects to people as parent and child. Two graphs G And H are ***equal*** if V(G) = V(H) and E(G) = E(H), in which case we write G = H.

Example 1:

Draw the graph corresponding to the vertex sets V = $\{v_1, v_2, v_3, v_4, v_5, v_6\}$ and edge sets E = $\{(v_1, v_2), (v_1, v_5), (v_1, v_6), (v_2, v_6), (v_3, v_4), (v_3, v_5), (v_4, v_5), (v_4, v_6)\ (v_5, v_6)\}$.

Solution

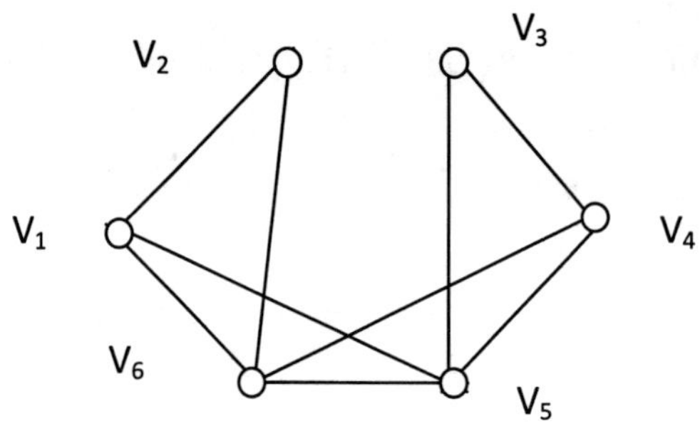

Figure 1.4

To make you comfortable with the basic idea of graph, one more example has been given below.

Example 2:

Let G = (V, E) where V = $\{v_1, v_2, v_3, v_4, v_5, v_6\}$, and edges E = $\{e_1, e_2, e_3, e_4, e_4, e_5\}$, and the ends of the edges are given by, $e_1 \leftrightarrow (v_1, v_4)$, $e_2 \leftrightarrow (v_1, v_6)$, $e_3 \leftrightarrow (v_2, v_5)$, $e_4 \leftrightarrow (v_4, v_5)$, $e_5 \leftrightarrow (v_5, v_6)$.

Solution

We can represent it graphically as in Figure 1. 5.

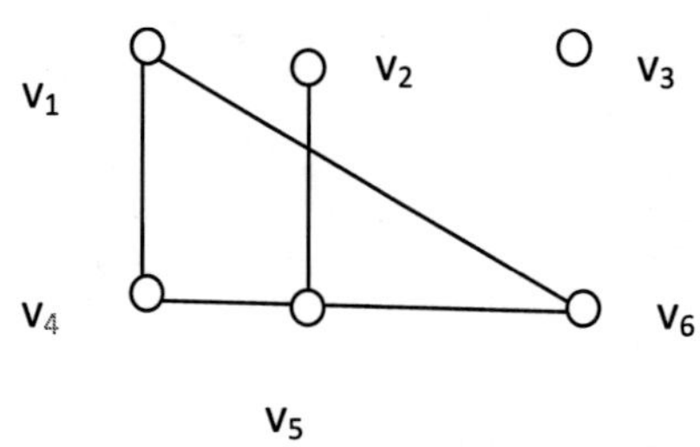

Figure 1.5 A graph with six vertices and five edges.

In drawing a graph, it is immaterial whether the lines are drawn straight or curved, long or short, what is important is the incidence between the edges and vertices.

The definition of the graph contains no reference to the length or the shape and positioning of the edge joining any pair of vertices, nor does it prescribe any ordering of positions of the vertices. Therefore, for a given graph, there is no unique diagram which represents the graph. We can obtain a variety of diagrams by locating the vertices in an arbitrary number of different positions and also by showing the edges by arcs or lines of different shapes. Because of this arbitrariness it can happen that two diagram's which look entirely different from one another may represent the same graph, because incidence between edges and vertices is the same in both cases. Generally a number of different diagrams may represent the same graph. For example,

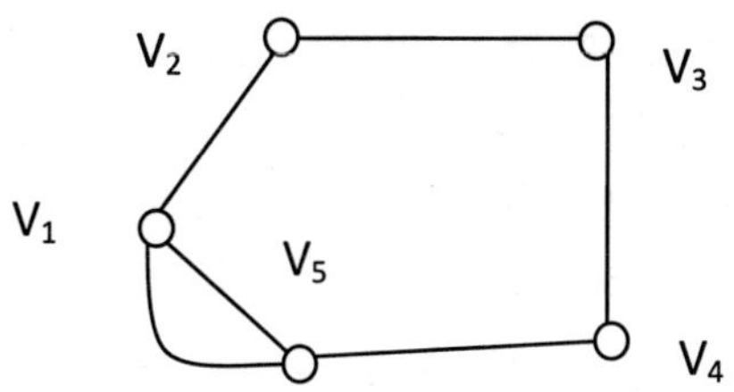

Figure 1.6 (a)

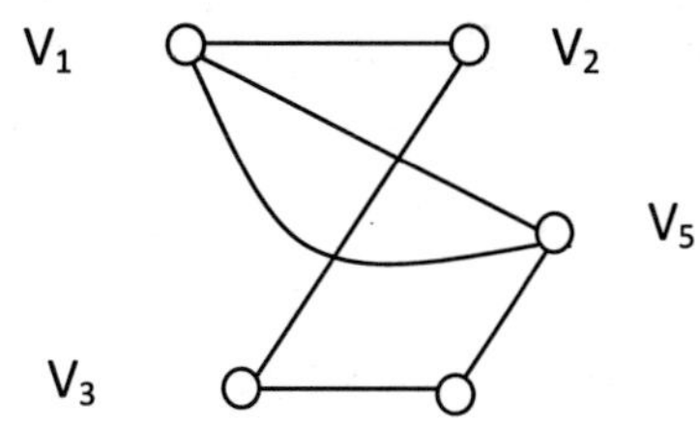

Figure 1.6 (b)

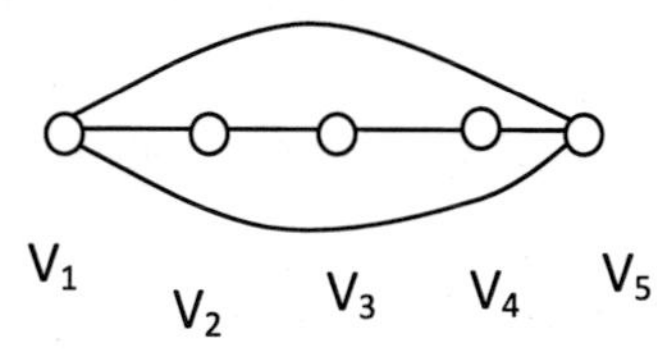

Figure 1.6 (c)

Figures 1.6 (b) and 1.6 (c) represent different drawings of the graph of figure 1.6 (a), with the vertex sets V = { v_1, v_2, v_3, v_4, v_5}, and edge sets E= {((v_1, v_2), (v_2, v_3), (v_3, v_4), (v_4, v_5), (v_5, v_1), (v_5, v_1)}, because incidence between edges and vertices is the same in both cases.

A graph in which every edge is directed is called **directed *graphs or simply digraphs***. Just as with graphs, digraphs have diagrammatic representation. A digraph is represented by a diagram of its underlying graph together with arrows on its edges, the arrow pointing toward the head of the corresponding arc. A digraph and its underlying graph are shown in Figure 1. 7.

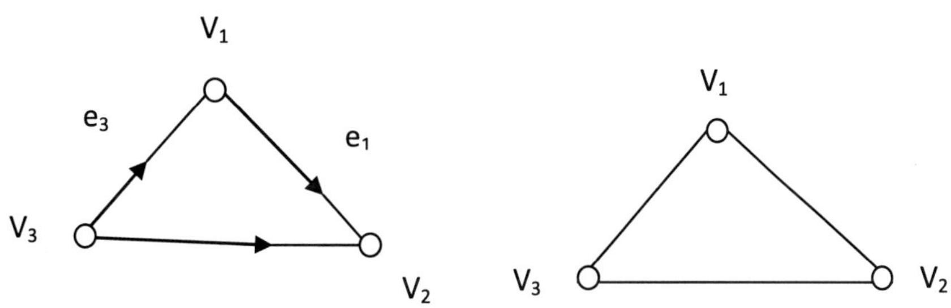

Figure 1.7 Digraph D and its underlying graph G

In directed graphs, edges have a direction (i.e., from one node to another). In undirected graphs, edges have no direction. Directed graphs are more appropriate for representing systems in which the direction of interaction is important (For example, in an ecological system, members of one species eat members of another species) while undirected graphs work better if the interactions has no specific direction. (i.e., symbiotic relation between two species in an ecological system- Of course this could also be seen as a pair of directed arcs between nodes representing two species). In Figure 1.8, an ecological system is presented schematically, where arrows are used to show the direction of interaction (Directed arcs represent Consumer – Food Relationship; with the arc being directed towards the food species). Thus, information can be represented as a graph with vertices and edges.

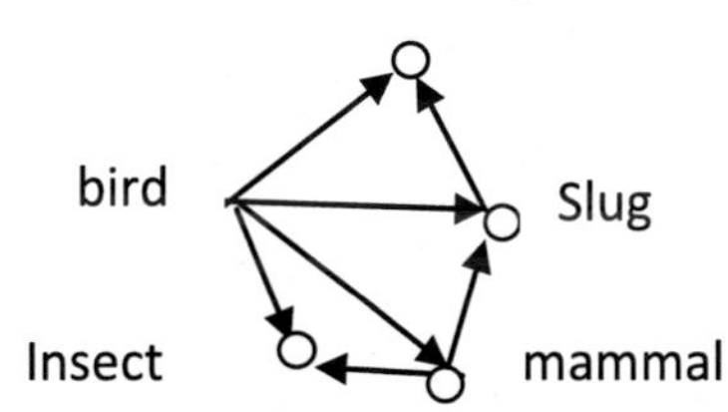

Figure 1.8 A simple Ecological system.

In the definition of graph, We usually disregard any direction of the edges and consider (u, v) and (v, u) as one and same edge in G (i.e., there is no distinction between the two vertices associated with each edge). In that case G is defined as an **undirected graph**. Also, it is possible for the edge set to be empty (Null Graph). Also the set of vertices V of a Graph may be infinite or finite. A graph with an infinite vertex is called an **infinite graph**. And in comparison, a graph with a finite vertex set (and edge set) is called a **finite graph**. In this book we will usually consider only finite graphs (for which V(G) is non empty finite set), and unless otherwise stated the term "graph" mean a finite graph.

To make the idea more clear, we cite a graph model for a network of holdings (herds) (Figure 1.9). The circles representing holdings, are labeled a through h, and the connection between them are labeled as number of animals transported in one day. Note, every holding does not have a transport (i.e., an isolated vertex h). In the graph theory terminology, each holding in the network is represented by a vertex and each transport by an edge. As specified earlier, A Graph consists of a vertex set V and an edge set E. Thus we write, G = (V, E). where V is the vertex set and E the edge set. The size of the vertex set (number of vertices) is expressed as |V| and size of the edge set as |E|. An edge is an ordered pair (u, v) consisting of vertices connected by the edge. The ordered pair (u, v) indicates the edge that connects the vertex u tovertex v. Thus for the holding net work of Figure 1.10, we have:

V = { a, b, c, d, e, f, g, h}

E = {(a, c), (b, d), (b, f), (c, b), (c, e), (c, g), (d, b), (d, e), (f, b), (f, g)}

G = (V, E).

If an edge (u, v) with u as source and v as the target vertex is distinct from the edge (v, u) the edge is directed. If the vertex ordering does not matter so that (u, v) & (v, u) are the same, the edge is undirected. A graph could either be directed or undirected, meaning that the edge set in the graph consists of respectively directed or undirected edges. If a group of vertices in an undirected graph are reachable from one another they are strongly connected. That is, strongly connected vertices are a group of vertices in a directed graph that are mutually reachable.

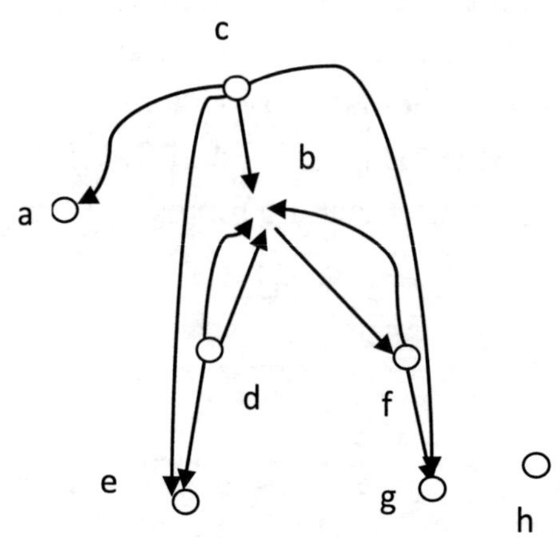

Figure 1.9 An example of a network of holdings, with the connections labeled with the number of animals transported per day.

Vertices are also sometimes called ***points or nodes***. The number of vertices in a graph G is called the ***order*** of G, while the number of edges is its size. Since the vertex set of every graph is non empty, the order of every graph is at least 1. The graph of Figure 1.4 has order 6 and size 9. We often use the terms n and m (when there is no explicit reference to the graph G) for the order and size respectively, of a graph. So for the graph G of Figure 1.4, n = 6 and m = 9. A graph with exactly one vertex (i.e., a graph with no edges) is called a ***trivial or Empty*** graph, implying that the order of a ***Non trivial*** graph is at least 2.

So far, we have explored graphs with vertices and edges listed explicitly. There are occasions when we are interested in the structure of the graph rather than explicitly listing its vertices and edges. In this case, (if the graphical representation is adequate for all discussions) a graph is drawn without labeling its vertices. A graph G is labeled when the n points are distinguished from one another by names such as $v_1, v_2, \ldots, v_n$. Figure 1.10 (a) shows a labeled graph and Figure 1.10 (b) an unlabelled graph.

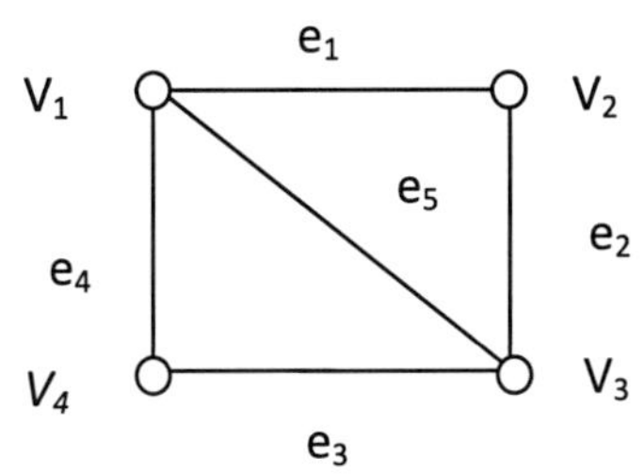

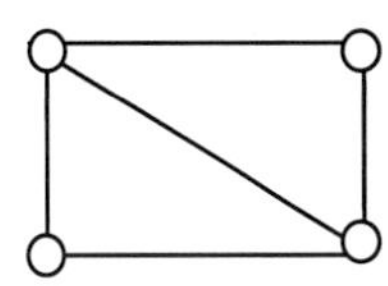

Figure 1.10 (a) A labeled graph **Figure 1.10 (b)** An unlabelled graph

It may so happen that, in a diagram of a graph, sometimes two edges may seem to intersect at a point that does not represent a vertex, for example edges e and f in Fig.1.11. Such edges should be thought of as being in different planes and thus having no common point.

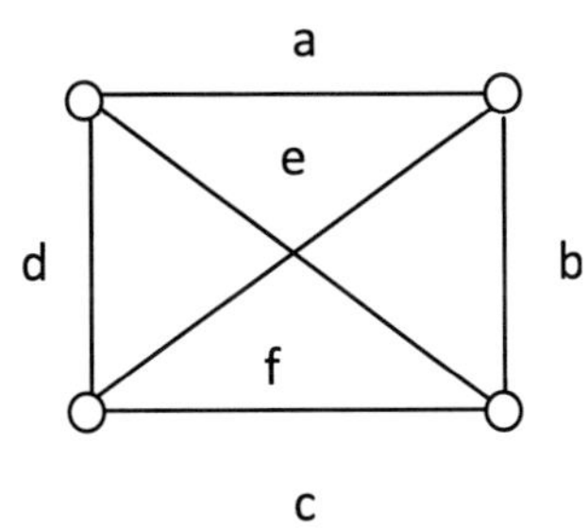

Figure 1.11 Edges e and f have no common point.

1.2 MORE DEFINITIONS

We hereby give some definitions to make you understand some of the basic concepts.

Parallel Edges: If two (or more) edges of a graph G have the same end vertices, then these edges are parallel. For example, the edges e_3 and e_4 of the graph of Figure 1.12 are parallel.

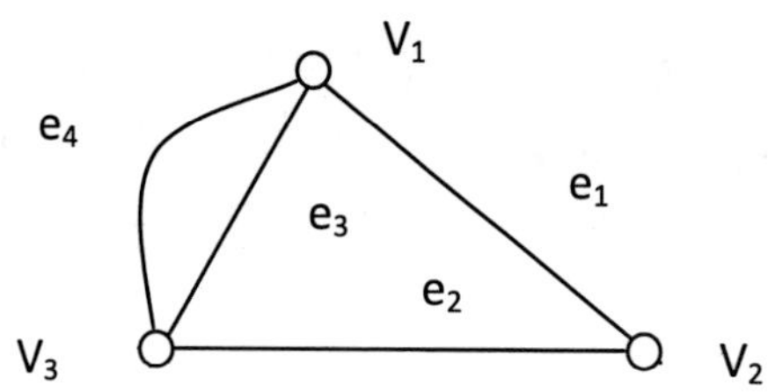

Figure 1.12

Isolated Vertex: A vertex of a graph G, which is not the end of any edge or of degree zero is called Isolated. For example, the vertex v_3 of Figure 1.5 is Isolated.

Neighbors or Adjacent Vertex: Two vertices which are joined by an edge are called adjacent or neighbors . The set of all such neighbors of vertex v is called the open Neighborhood of v and it is denoted by N(v); the set N [v] =N (v) U {v} is the closed neighborhood of v in G. When G must be explicit, these open and closed neighborhoods are denoted by N_G (v) and N_G [v], respectively. For example, in the graph of Figure 1.12, vertices v_2 and v_3 are adjacent. The neighborhood set $N(v_2)$ is $\{v_1, v_3\}$, $N(v_3) = \{ v_2, v_1\}$, $N[v_2] = \{v_1, v_3, v_2\}$ and $N[v_2] = N(v_2)$ U v_2. Further, v_1 and v_2 are adjacent vertices, and e_1 and e_2 adjacent edges.

Incidence: An edge e of a graph is said to be incident with the vertex v if v is an end vertex of e (or v is incident with e). Two edges e and f which are incident with a common vertex v are said to be adjacent.

It is natural to count the edges that are incident with a particular vertex. i.e., Given a vertex v, we can find the number of edges that are incident with v. If e = {u, v}, where $u \neq v$ is an edge, then e will be counted once while counting the edges that are incident with u, and again it will be counted once while counting the edges that are incident with v, with this in mind, we make a convention that a loop e will be counted twice when finding the number of edges that are incident with u.

Next we will give a name to the number obtained by counting all the edges that are incident with a vertex v.

Degree: The degree d_G (v) (orvalency) of any vertex v of a graph G is the number of edges of G incident with v. The d_G (v) can also be denoted by deg_G (v) (or explicitly, we use d(v) or deg (v)) to denote the degree of the graph. Also, d(v) is the set of neighbors of a vertex (or number of vertices adjacent to v) . Thus, d (v) = |N (v)|. Each loop is counted twice or it is the number of times v is an end vertex of an edge. A vertex of degree zero is ***isolated***. It follows that an isolated vertex is not adjacent to any vertex and a graph with only isolated vertices is called a ***null graph***. Moreover, a vertex of degree 1 is a pendant (or an ***end-vertex*** or a ***leaf vertex)*** vertex. Consequently, a pendent vertex is adjacent to exactly one other vertex. Vertex v_2 in the graph of Figure 1.5 is pendant. For example, consider the graph G of Figure 1.13.

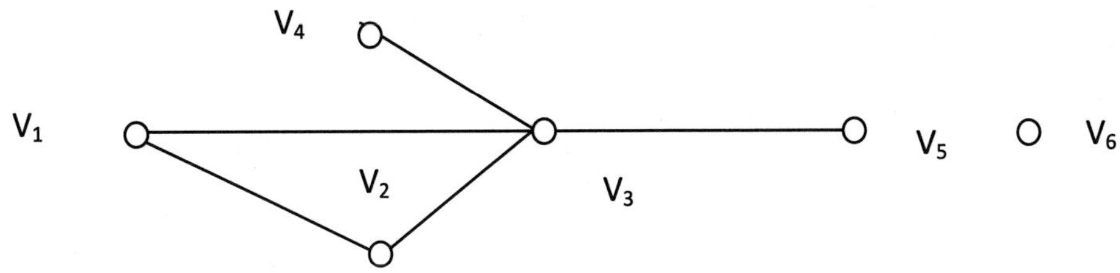

Figure 1.13 A graph G with with $\delta(G)=0$ and $\Delta(G)=4$

The graph in Figure 1.13, has order six vertices (order 6) and five edges (size 5). Each vertex of the graph G is labeled by its degree. i.e., $d(v_1)= deg(v_2) = 2$, $d(v_4) = d(v_5) = 1$, $d(v_3) = 4$, $d(v_6) = 0$. The minimum of all the degrees of the vertices of a graph G is denoted by $\delta(G)$ and the maximum of all the degrees of the vertices of G is denoted by $\Delta(G)$. Since G contains an isolated vertex namely v_6, it follows that $\delta(G) = 0$. Further more, v_3 has the largest degree in G. So, $\Delta(G) = 4 = d(v_3)$. Both v_4 and v_5 are end vertices. (i.e., $d(v_4) = d(v_5) = 1$). So if a graph is of order n and v is any vertex of G, then $0 \leq \delta(G) \leq deg(v) \leq \Delta(G) \leq n-1$. On the other hand, If $\delta(G) = \Delta(G) \leq k$, that is, if all the vertices have the same degree k, then it is k-regular. A 3-regular graph is Cubic. The graphs K_4, $K_{3,3}$, Q_3 are cubic graphs. However, the best known cubic graph may very well be the Petersen Graph (Q_3), (Figure 1.14).

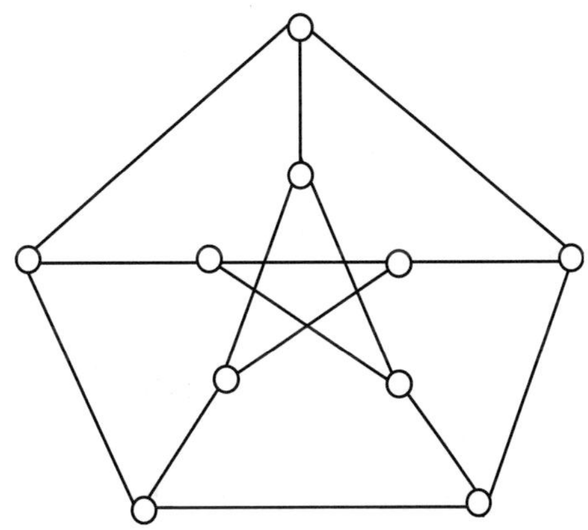

Figure 1.14 The Petersen graph Q_3

The concept of degree has counterparts in both multigraphs and digraphs. For a vertex v in a multigraphs G, the degree deg(v) of v in G is the number of edges of G incident with v, where there is contribution of 2 for each loop at v. For the multigraphs G of Figure 1.15(a),

$$deg(u_1) = 4, deg(u_2) = deg(u_3) = 6, deg(u_4) = 4$$

For a vertex v in a digraph D, the **out degree** od v of v is the number of vertices of D to which v is adjacent, while the **in degree** idv of v is the number of vertices of D from which v is adjacent. For the digraph D of Figure 1.15 (b), $odv_1 = idv_1 = 1$, $odv_2 = 2$, $idv_2 = 1$, $odv_3 = 0$, $idv_3 = 1$.

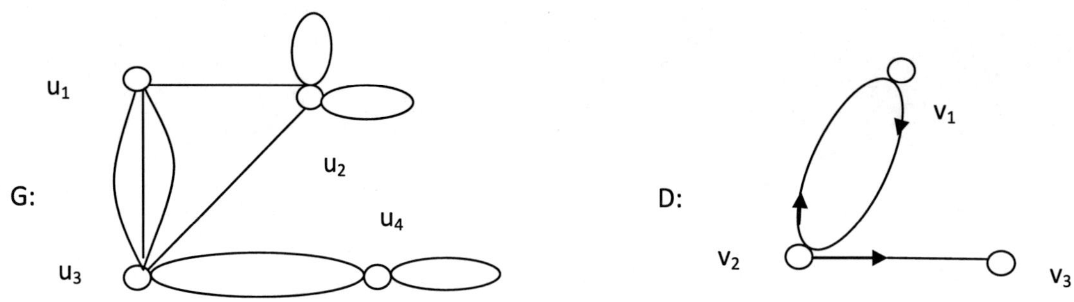

Figure 1.15 Illustrating degrees in a multigraph and a digraph.

There is a great deal of information that can be learned about a graph from the degree of its vertices. Now, What do we get when we add the degree of all the vertices of a graph G = (V, E)? Each edge contributes two to the sum of the degrees of the vertices because an edge is incident with exactly two (possibly equal) vertices. This means that the sum of the degrees of the vertices is twice the number of edges. We thus have a result in Theorem 1.1, due to Euler (1707-1783), which was the first theorem of graph theory, which is sometimes called the ***Handshaking Theorem***, because of the analogy between an edge having two end points and a handshake involving two hands. This theorem connects the degrees of the vertices and the number of edges of a graph.

THEOREM 1.1 THE HANDSHAKING THEOREM

For any graph G with e edges and n vertices $v_{1,} ...., v_n$

$$\sum_{i=1}^{n} d(v_i) = 2e \tag{1.1}$$

Proof: Since degree of a vertex v in a graph G is the number of edges connected with it, with loops being counted twice, the sum of the degree counts the total number of times an edge is incident (connected) with a vertex v. As every edge is connected with exactly two vertices, when summing the degree of the vertices of a graph G, each edge is counted twice at each of its end, one for each of the two vertices incident with the edge. This implies that the sum of the vertex degrees is equal to twice the number of edges. The total degree of a graph is equal to two times the number of edges, with of loops included.

Taking Figure 1.16 as an example, the graph has eight vertices with each vertex having a degree of three. Since $\sum_{i=1}^{n} d(v)=2e$. We have 3(8) = 24 = 2 × e . It must have 12 edges, and it does.

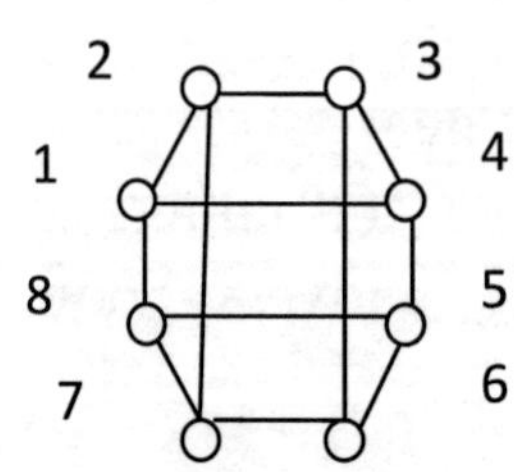

Figure 1.16

Example 3:

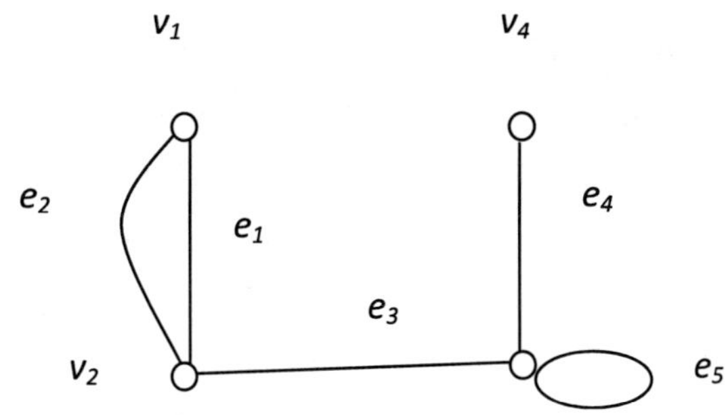

Figure 1.17 A Pseudo graph with four vertices and five edges

The Pseudo graph in Figure 1.17, (A Pseudo graph is like a graph, but it may contain loops and /or multiple edges) has vertices of degree 4, 3, 2, 1. Since 4 + 3 + 2 + 1 = 10, the graph must have five edges and it does. (Note that a loop is one edge, but it adds two to the degree.)

Odd or Even Vertex: A vertex of a graph is called odd or even depending on whether its degree is even or odd. Returning to the graph of Figure 1.13 we see that it has two odd vertices v_4 and v_5 three even vertices v_1, v_2 and v_3. In particular, the number of odd vertices of G is even. We show that this is the case for every graph. This simple fact has many consequences, one of which is given as Theorem 1.2.

THEOREM 1.2

The number of vertices of odd degree in a graph is always even.

Proof: If we consider the vertices with odd and even degrees separately, the quantity on the left side of Eqn. (1.1) can be expressed as the sum of two sums, each taken over vertices of even and odd degrees, respectively, as follows:

$$\sum_{i=1}^{n} d(v_i) = \sum_{even} d(v_i) + \sum_{odd} d(v_k) \tag{1.2}$$

By the previous theorem. $\sum_{i=1}^{n} d(v_i) = 2e$, *an even number.* Since the left hand side in Eqn. (1.2) is even, and the first expression on the right hand side is even(being the sum of even numbers), the second expression must also be even.

$$\sum_{odd} d(v_k) = \text{an even number} \tag{1.3}$$

Because all the terms in this sum are odd, there must be even number of such terms to make the sum an even number.. Hence the theorem.

1.3 DEGREE SEQUENCES

Although we have been discussing graphs all of whose vertices have the same degree, it is more typical for the vertices of a graph to have a variety of degrees. A sequence formed by the degrees of vertices of G is called a ***degree sequence*** of G. Furthermore, If $v_1, v_2, \ldots\ldots\ldots v_n$ are the vertices of G, then the sequence ($d_1, d_2, \ldots\ldots\ldots, d_n$), where d_i = degree (v_i), is the ***degree sequence*** of G. It is customary to give this sequence in the non increasing or non decreasing order. Usually, we order the vertices so that the degree sequence is monotone increasing, that is, so that $\delta(G) = d_1 \leq d_2 \leq \ldots\ldots. \leq d_n = \Delta(G)$).

Determining a degree sequence of a graph is not difficult. There is a converse question, " Does a given a degree sequence has an underlying simple graph" - that is considerably more intriguing.

The degree sequence d = ($d_1, d_2, \ldots d_n$) is graphic if there is a simple undirected graph with degree sequence d.

There are potential difficulties in determiningthe sequences, graphical or not . An efficient theorem that will help us to determine which sequence is graphical, is due to Vaclav Havel and S. Louis Hakimi. To use this theorem, we assume that we are beginning with a non-increasing sequence.

THEOREM 1.3 (HAVEL HAKIMI): (WITHOUT PROOF)

A non-increasing sequence S = $d_1, d_2, \ldots\ldots d_n (n \geq 2)$ of non-negative integers, where $d_1 \geq 1$, is graphical if and only if the sequence

$$S_1 = d_2 -1, d_3 -1, \ldots\ldots., d_{d\,i+1} -1, d_{d^i+2}, \ldots, d_n \text{ is graphical}$$

For example the sequence (6, 6, 6, 6, 4, 3, 3, 0) is not graphical. By using Havel Hakimi Theorem for the sequence: That is, first we take the first term of the sequence namely 6, Eliminate the first term 6 and reduce the next 6 terms by one, we get the sequence 5, 5, 5, 3, 2, 2, 0. This sequence is in descending order. The first term of the sequence is 5, thus eliminating the first term and reducing the numbers in the next 5 terms by one, we get the sequence 4, 4, 2, 1.1.0. The first term of the sequence is 4, thus eliminating the first term and reducing the numbers in the next 4 terms by one, we get the sequence 3, 1, 0, 0, 0. There exists no graph having one vertex of degree 3 and other vertex of degree1. Therefore, the last sequence is not graphical. Hence the given sequence is also not graphical.

Example: 4

Which of the following sequences are graphical ?

(1) d = 6, 5, 5, 4, 3, 3, 3, 2, 2.

(2) d = 2, 2, 3, 5.

(3) d = 6, 5, 5, 4, 3, 3, 2, 2, 2.

Solution

(1) It cannot be a degree sequence of a graph and hence not graphical, since it has an odd number of terms that are odd integers.

(2) Graphical and the graph with this degree sequence is shown in Figure.

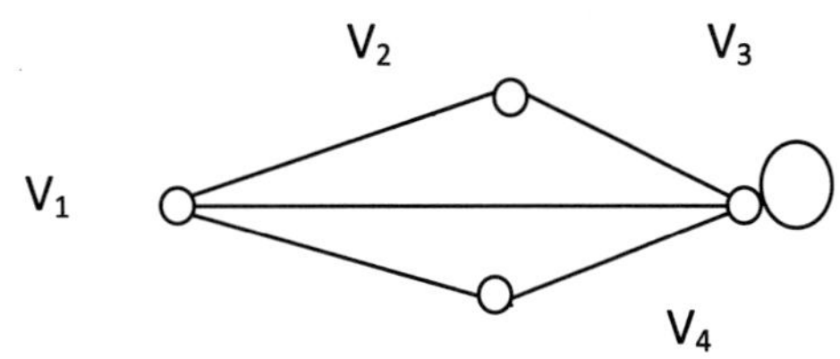

Figure 1.18 A graph with degree sequence 2, 2, 3, 5

(3) Using Havel Hakims theorem, the given sequence is graphical. Reducing the sequence as follows:

6 5 5 4 3 3 2 2 2

4 4 3 2 2 1 2 2

4 4 3 2 2 2 2 1

3 2 1 1 2 2 1

3 2 2 2 1 1 1

1 1 1 1 1 1

The Sequence (1 1 1 1 1 1) is graphical.

1.4 TYPES OF GRAPHS

There are several variations of graphs which deserve mention. Whenever it is necessary to draw a strict distinction, it may be useful to define the term graph with different degrees of generality. Most commonly, in modern texts in graph theory, unless otherwise stated, graph means "Undirected simple finite graphs".

Undirected Graph: A graph in which edges have no orientation, i.e., they are not ordered pairs, but sets {u, v} (or 2-multisets) of vertices. In an undirected graph we can refer to an arc joining the vertex pair u and v as either (v, u) or (u, v). An undirected graph is also defined in the same manner as directed graph except that edges (arcs) are unordered pairs of distinct vertices. An undirected arc (u, v) can be considered as a two-way road with traffic flow permitted in both directions: either from vertex u to v or from v to u. An edge such as {u, v} stands for {(u, v), (v, u)}. Although (u, v) = (v, u) only when u = v.

Mixed Graph: A graph G in which some edges may be directed and some may be undirected. It is written as an ordered triple G = (V, E,) with V, E, and defined as above. Directed and undirected graphs are special cases. The mixed graph M is called ***simple*** if it has no loops, and no parallel edges.

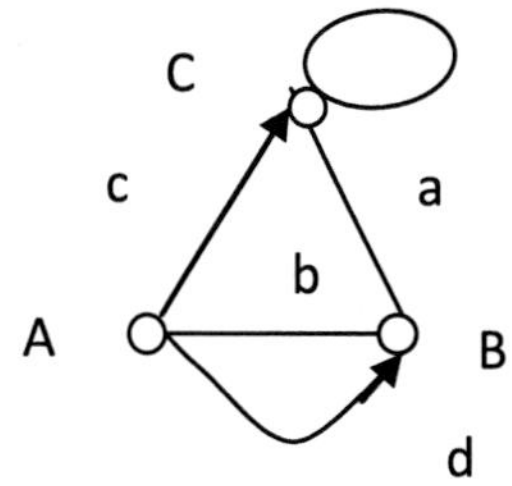

Figure 1.19 Mixed graph

Multigraph: It is a graph in which there are multiple edges (also called "parallel edges") between a pair of vertices. Formally, a multigraphs G is an ordered pair G = (V, E) with V -a set of vertices or nodes and E a multi set of unordered pairs of vertices called edges or lines. Or in multigraphs, no loops are allowed but more than one line can join two points. If both loops and multiple lines are permitted, we have a ***Pseudo graph***. Figure 1.20 shows a Multigraph and a Pseudo graph with the same underlying graph, a "***triangle***".

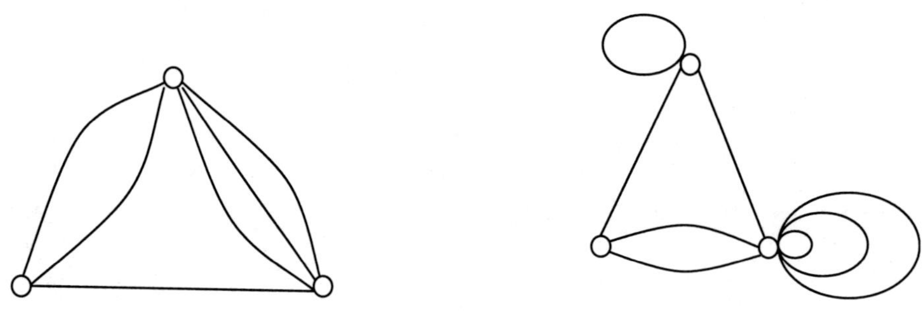

Figure 1.20 A Multigraph and a Pseudo graph

Note 1: Every simple and Multigraphs is a Pseudo graph but the converse is not true.

Directed Graph: As already stated, a directed graph or Digraph D is a graph each of whose edge is directed. A directed edge is an edge such that one vertex incident with it is designated as the head vertex and the other incident vertex is designated as the tail vertex. In this situation, we may assume that the set of edges is subset of the ordered pairs V × V. Or in other words, A digraph D consists of a finite non empty set V of points together with a prescribed collection X of ordered pairs of distinct points. The elements of X are directed lines or arcs. A directed edge uv is said to be directed from its tail u to its head v. By definition, a digraph has no loops or multiple arcs (Figure 1.21.).

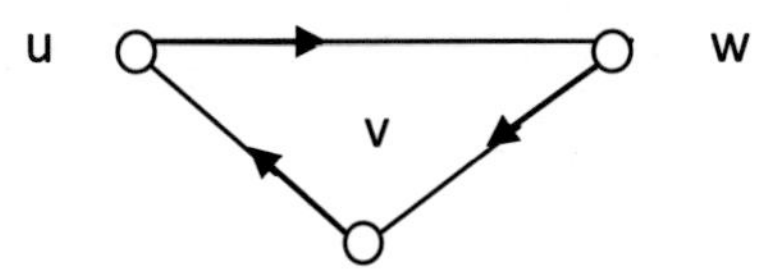

Figure 1.21 Digraph D

The in degree id(v) of a vertex v ∈ V(G) count the number of edges such that v is the head of those edges (i.e., number of edges ending from it). The out degree od(v) of a vertex v ∈ V(G) counts the number of edges such that v is the tail of these edges (i.e., number of edges beginning from it). Self loop at a vertex contributes 1 to both in-degree and out-degree of this vertex. The degree deg(v) of a vertex is the sum of in degree and out degree of v.

A vertex of D is isolated if its degree is 1. Thus for a pendent vertex v, either $d^+(v) = 1$ and $d^-(v) = 0$, or $d^+(v) = 0$ and $d^-(v) = 1$.

Example: 5

For the graph of Figure 1.22 find the in degrees and out degrees of various vertices.

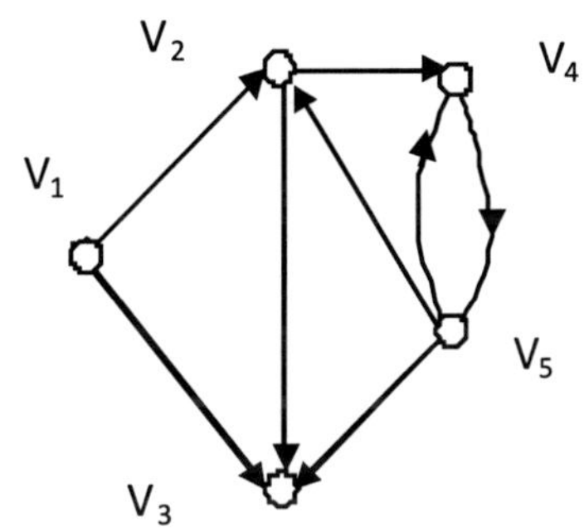

Figure 1.22 Directed graph

Solution

For the graph of Figure 1.22, the in degrees and out degrees of various vertices are as index.

Vertex	v_1	v_2	v_3	v_4	v_5
In degree	0	2	3	2	1
Out degree	2	2	0	1	3

THEOREM 1.4

If G(V, E) is a directed graph, then the sum of the out degrees of the vertices of the graph equals the sum of in degrees of the vertices which equals the number of edges in a digraph.

i.e., $\sum indegree(v) = \sum outdegree(v) = e, \forall\, v \in V$

Proof *:* The proof of the theorem is obvious. As shown in Figure 1. 22, the sum of the in degrees equals the sum of out degrees which also equals the number of edges in the digraph.

A directed graph D is called ***symmetric*** if, for every arc in D, the corresponding inverted arcs are also belong to D. A symmetric loop less directed graph D = (V, A) is equivalent to a simple undirected graph G = (V, E), where the pairs of inverse arcs in A correspond to one-to-one with the edges in E, thus the edges in G number |E| = |A| / 2 or half the number of arcs in D (Figure 1.23(a)). A digraph which is both simple and symmetric is called a simple symmetric graph. A digraph that has almost one directed edge between a pair of vertices (loops are allowed) is called ***asymmetric graph***. (Figure 1.23(b)). A digraph which is not simple and asymmetric is called a simple asymmetric graph.

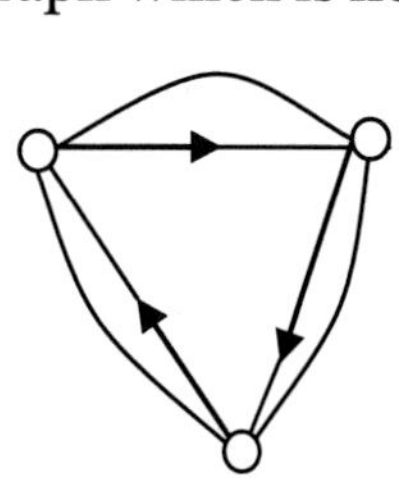

Figure 1.23 (a)

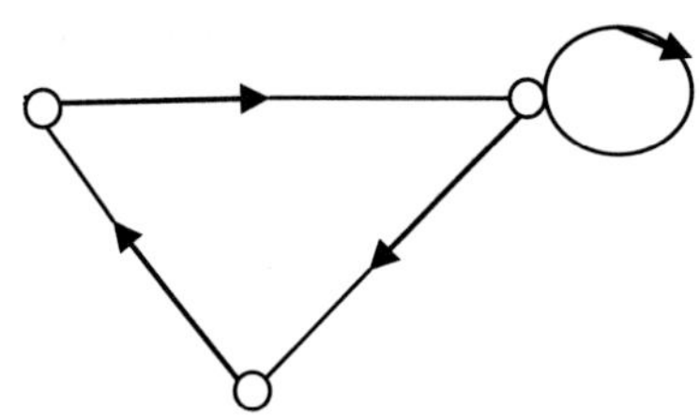

Figure 1.23 (b)

The undirected graph that results from ignoring directions of edges is called underlying undirected graph (Figure 1.24). A graph with directed edges and its underlying graph have the same number of edges. Thus, every digraph D defines a unique graph G. Conversely, given any graph G, we can obtain a digraph D from G by specifying for each edge of G. Such a specification is called an ***orientation*** of G. A digraph is ***strict*** if its underlying graph is simple.

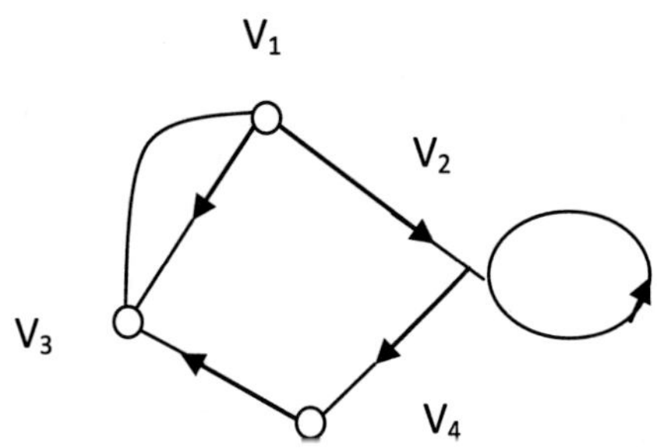

(a) Digraph D

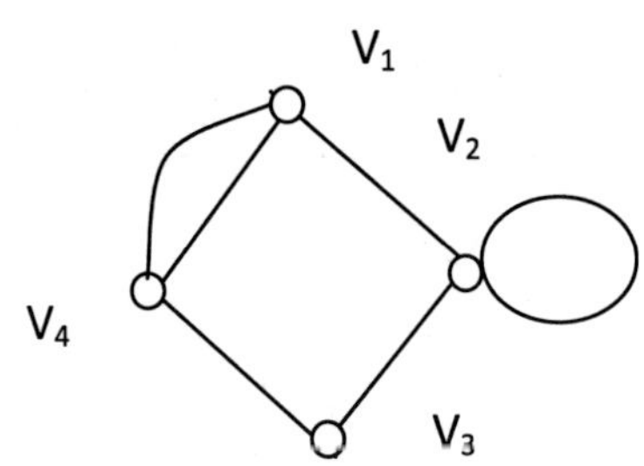

(b) Underlying undirected graph G of D

Figure 1.24

A vertex v is reachable from a vertex u of d if there is a directed path in d from u to v. Two vertices of d are ***diconnected*** if each is reachable from the other. clearly, diconnection is an equivalence relation on the vertex set of d and if the equivalence classes are v_1, v_2, … v_n, the sub graphs of d induced by v_1, v_2, …. , v_n are called the *dicomponents* of d. A digraph is diconnected (strongly connected) if it has exactly one dicomponents. A diconnected digraph is also called a strong digraph.

Reversal Graph*:** The converse of a directed graph G (V, E) is a directed graph denoted by G'(V, E'), Where E' is the converse of E. The diagram of G' is obtained from G by simply reversing the directions of the edges in G. The converse G' is also called ***reversal graph (or) directional dual graph of G (Figure 1.25).

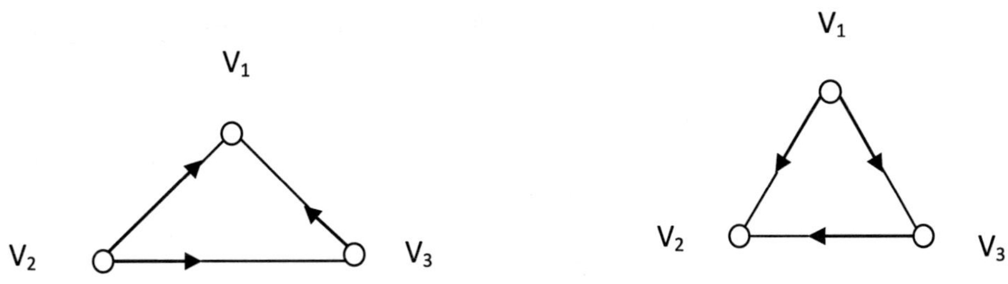

(a) A directed Graph G **(b)** Directional dual graph of G

Figure 1.25

Oriented Graphs*:** A variation of the definition of digraph, is the ***oriented graph, in which not more than one of (x, y) and (y, x) may be arcs. An oriented graph D is obtained by assigning a direction to each edge of some graph G. The digraph D is also called an orientation of G. Thus, an oriented graph is a graph having no symmetric pair of directed lines. Figure 1.26 shows two graphs D_1 and D_2, where D_2 is an oriented graph but D_1 is not.

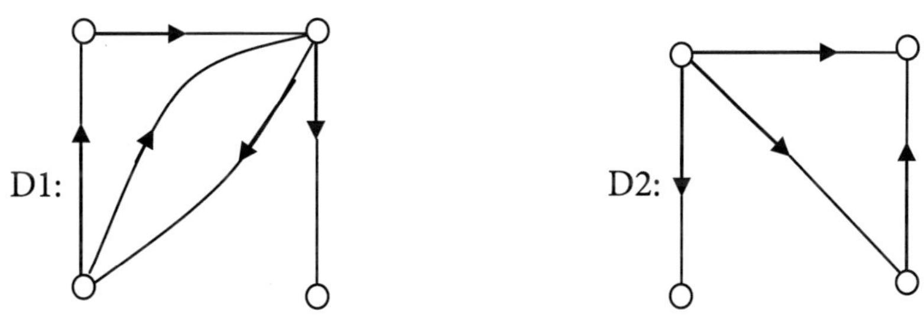

Figure 1.26 Digraphs: **(a)**Graph D_1 **(b)**Graph D_2 (Oriented Graph)

Simple Graph*:** A ***simple graph is an undirected graph that has no loops and no more than one edge (multiple edges) between any two different vertices (Figure 1.27). In a simple graph the edge of the graph form a set (rather than a multiset) and each edge is a pair of distinct vertices. In a simple graph with n vertices every vertex has a degree that is less than n (the converse however is not true- there exist non simple graphs with n vertices in which every vertex has a degree smaller than n). A simple graph with n vertices can have at most n (n-1) / 2 edges. K_n (complete graph with n vertices) has the maximum number of edges among all simple graphs with n vertices. On the other extreme, a graph may possess no edge at all. Such a graph is a totally disconnected graph. Thus, for a simple graph of size m and with n vertices, we have $0 \le m(G) \le n(n-1)/2$. Also $0 \le \deg v \le n-1$. Graphs are not necessarily simple, unless we explicitly say so.

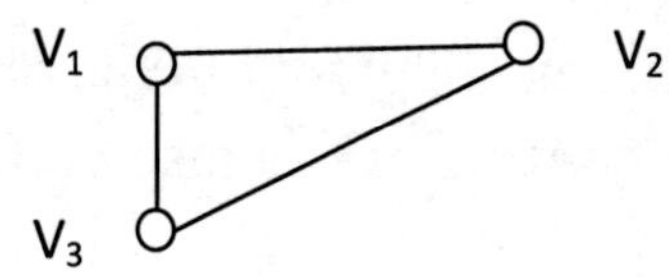

Figure 1.27 A simple graph without loops and parallel edges.

Null (or Trivial) Graph*:** In the definition of a graph G = (V, E), it is possible for the edge set E to be empty. Such a graph, without any edges, is called a ***null (or empty)graph (Figure 1.28). In other words, every vertex in a null graph is an isolated vertex. Although edge set E may be empty, the vertex set V must not be empty; otherwise, there is no graph. In other words, by definition, a graph must have at least one vertex. A null graph of order n is a graph with n vertices and no edges denoted by N_n.

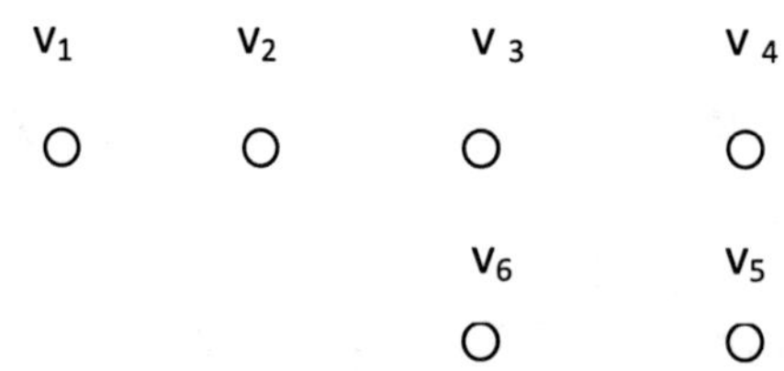

Figure 1.28 Null graph with six vertices

Weighted Graph*:** A graph is a ***weighted graph if a number is assigned to each edge (Figure 1.29). Such weights must represent, for example, costs, lengths, or capacities, etc., depending on the problem. The weight of the graph is the sum of the weights given to all edges, showing how strongly the elements effect one another. The weight is a number that expresses the influence of one element to the other one and it is positive or negative depending on whether this effect acts positively or negatively on the element. That is, we obtain a weighted directed graph. For the graph of Figure 1.8 representing ecological system, the plus one weighted link is denoted by the direction of the arrow, and the -1 weighted link by the opposite direction, eg., a bird feeds on an insect and an insect is consumed by bird.

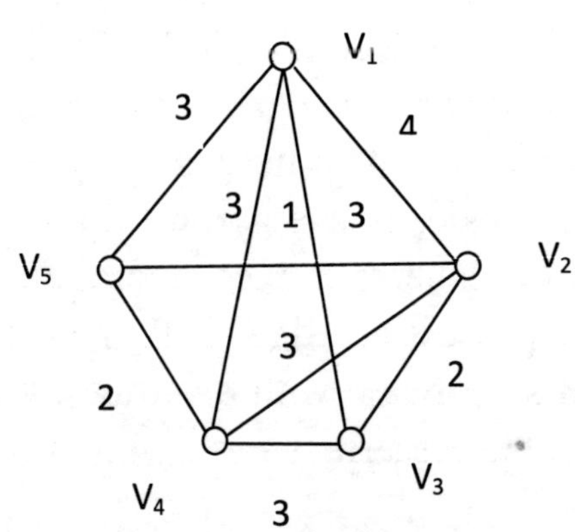

Figure 1.29 A weighted graph with five vertices

Regular Graph*:** A ***regular graph is a graph where each vertex has the same number of neighbours, i.e. every vertex has the same degree and valency. A regular graph with vertices of degree k are called k- regular graph or regular graph of degree k. (Figure 1.30). Regular graphs of degree at most two are easy to classify: A 0-regular graph consists of disconnected vertices, a 1-regular graph consists of disconnected edges, and a 2-regular graph consists of disconnected cycles. A 3-regular graph is known as a cubic graph. i.e., Every null graph is a regular of degree 0 and a complete graph K_n is regular of degree n-1. There is no 1-regular or 3-regular graph of order 5. Note that if a graph G has n vertices and is regular of degree r, then it has (1/2) nr edges.

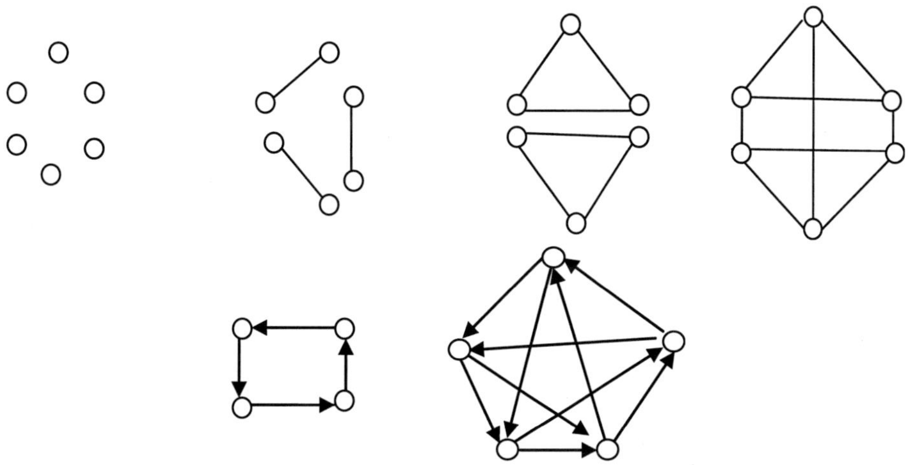

Figure 1.30 0-regular 1-regular 2-regular 3-regular undirected graphs and 1 regular & 2-regular Digraphs

A strongly regular graph on the other hand, is a regular graph where every adjacent pairs of vertices has the same number n of neighbours in common, and every non adjacent pairs of vertices has the same number n of neighbours in common. The smallest graph that are regular but not strongly regular are the cycle graph and the circulant graph on six vertices. A regular directed graph must also satisfy the stronger condition that the in degree and out degree of each vertex are equal to each other. A digraph D is regular of degree r, if od v = id v = r, for every vertex v of D.

***Complete Graphs*:** A complete graph K_n is a simple graph in which each pair of distinct vertices is joined by an edge. That is, every two distinct vertices are adjacent. The complete graphs on 1 through 5 vertices are shown in Figure 1.31. The complete graph of n vertices has n (n-1) / 2 edges. and is denoted by K_n. i.e. The complete graph K_n is strongly regular for any n. It is a regular graph of degree (n-1). The complete bipartite graph $K_{n,n}$ on 2n vertices is (n-1) regular

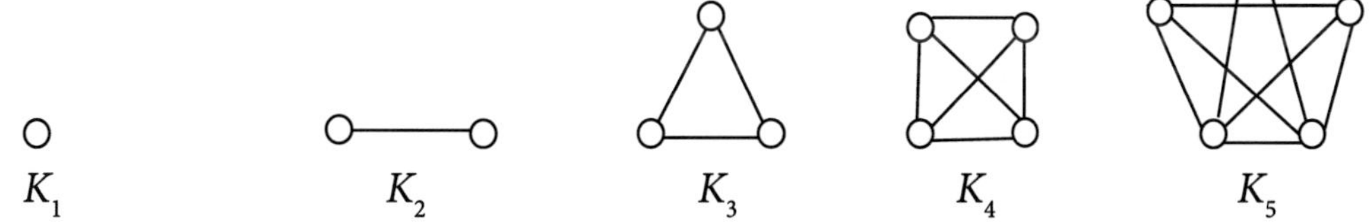

Figure 1.31 The First Five Complete Graphs.

Bipartite Graphs: A bipartite graph G is a graph whose vertex set V can be partitioned into two non-empty subsets X and Y in such a way that each edge of G has one end in X and one end in Y (i.e., every edge of G joins X with Y). The partition $V = X \upsilon Y$ is called bipartition on G. If G contains every edge joining X and Y, (i.e., every vertex in X is joined to every vertex of Y), then we call it a complete bipartite graph denoted by $K_{m, n}$, where m and n are vertices in X and Y respectively. The one vertex graph K_1 shown in Figure 1.31 is bipartite since bipartition sets are not required to be non-empty. It is not always easy to tell at a glance whether a graph is biparite.

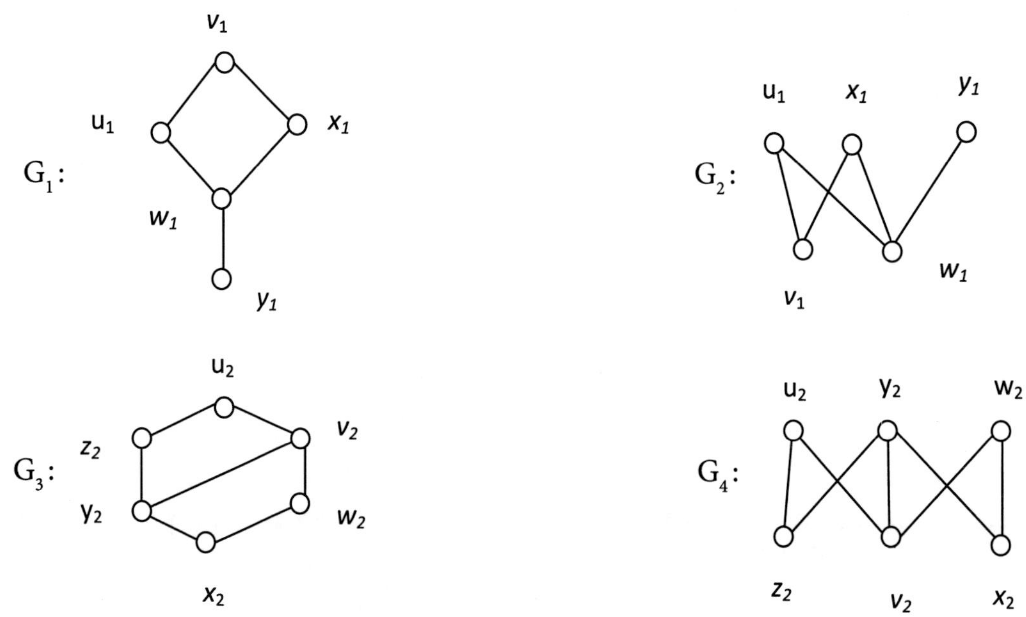

Figure 1.32 Bipartite graphs

For example, the connected graph G_1 and G_2 of Figure 1.32 are bipartite, as every edge of G_1 joins to a vertex of $u_1 = \{ u_1, x_1, y_1\}$ and every vertex of $W_1 = \{v_1, w_1\}$ while every edge of G_2 joins a vertex of $U_2 = \{ u_2, w_2, y_2\}$ and a vertex of $W_2 = \{v_2, x_2, z_2\}$. By letting $U = U_1 \cup U_2$ and $W = W_1 \cup W_2$, we see that every edge of $G = G_1 \cup G_2$ joins a vertex of U and a vertex of W. This illustrates the observation that a graph is bipartite if and only if each of its components are bipartite.

Certainly, not every graph is bipartite. For example, consider the 5-cycle C_5 in Figure 1.33. If C_5 were bipartite, then its vertex set can be partitioned into two sets U and W such that every edge of C_5 joins a vertex of U and a vertex of W . The vertex v_1 must belong to either U or W, say $v_1 \in U$. Since, $v_1 \in v_2$ is an edge of C_5, it follows that $v_2 \in W$. Since $v_2 v_3$ is an edge of C_5 it follows that $v_3 \in U$. Similarly, $v_4 \in W$ and $v_5 \in U$. However, $v_1 v_5 \in U$ and $v_1 v_5$ is an edge of C_5. This is a contradiction. Therefore, C_5 is not bipartite. In fact, no odd cycle is bipartite. Indeed, any graph that contains an odd cycle is not bipartite. The converse is true as well.

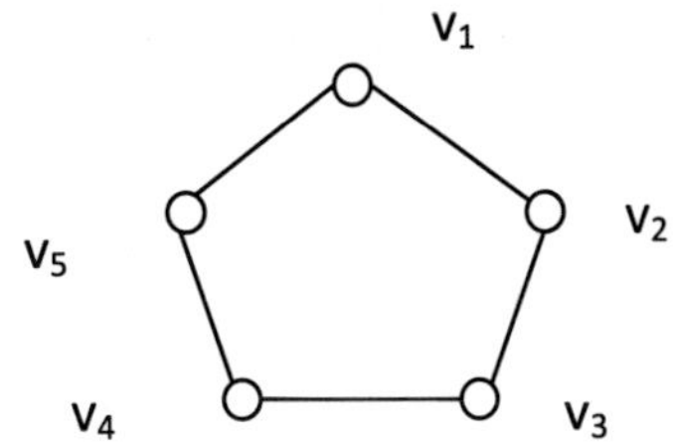

Figure 1.33 A 5 cycle. (A graph that is not bipartite)

The ***complete bipartite graph*** $K_{m,n}$, on the other hand, is a graph that has its vertex set partitioned into two subsets of m and n vertices, respectively. There is an edge between two vertices if and only if one vertex is in the first subset and the other vertex is in the second subset. If either m = 1 or n = 1, then $K_{m, n}$ is a star. Several complete bipartite graphs are shown in Figure 1.34. A complete bipartite graph $K_{m, n}$ has m + n vertices and mn edges. A star is a complete bipartite graph $K_{1, n}$. Clearly $K_{m,n}$ has mn lines.

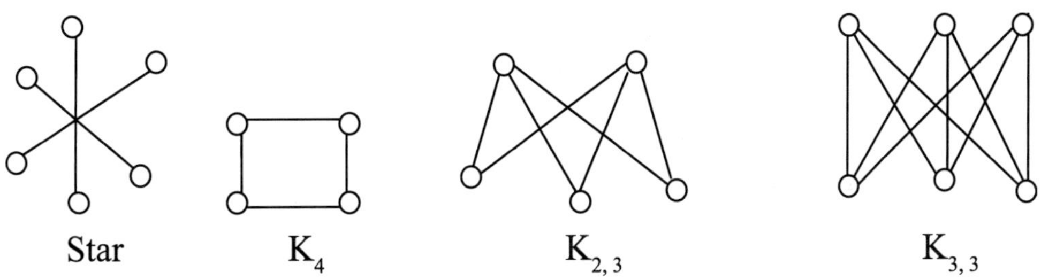

Figure 1.34 Some complete Bipartite Graphs

THEOREM 1.5

A graph is bipartite if and only if all its cycles are even.

Proof: If G is a bipartite graph, then its vertex set V can be partitioned into two sets X and Y. So that every edge of G joins a point of X with a point of Y. Thus every cycle $v_1v_2......v_nv_1$ in G necessarily has its oddly subscripted points in X, say, and others in Y, So that its length n is even. Hence the theorem.

Cycle graphs: A cycle graph of order 'n' is a connected graph whose edges form a cycle of length 'n' and denoted by C_n. The cycles C_3, C_4, C_5, and C_6 are displayed in Figure 1.35.

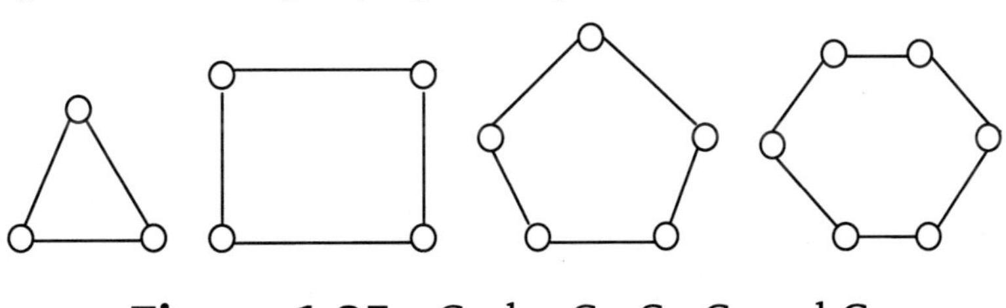

Figure 1.35 Cycles C_3, C_4, C_5 and C_6

Path Graph: A path graph of order 'n' is obtained by removing on edge from a C_n graph, denoted by P_n.

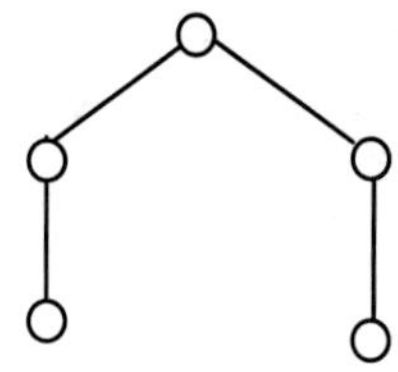

Figure 1.36 The path graph

Wheel Graph: We obtain the wheel W_n when we add an additional vertex to the cycle C_n, form ≥ 3, and connect this new vertex to each of the n vertices in C_n, by new edges. The wheels W_3, W_4, W_5 and W_6 are shown in Figure 1.37.

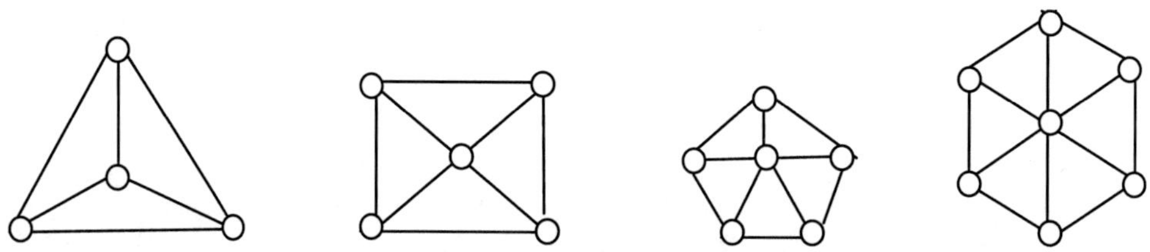

Figure 1.37 The Wheels W_3, W_4, W_5 and W_6

1.5 OPERATIONS ON GRAPHS

In mathematics, one always tries to get new structures from given ones. This also applies to the realm of graphs, where one can generate many new graphs from a given set of graphs. There are several ways to produce a new graph from a given pair of graphs. In this section we consider some of the methods of generating new graphs from a given pair of graphs.

Union: Given two graphs G_1 and G_2 their ***union*** (Figure 1.38) will be a graph such that $V(G_1 \cup G_2) = V(G_1) \cup V(G_2)$ and $E(G_1 \cup G_2) = E(G_1) \cup E(G_2)$.

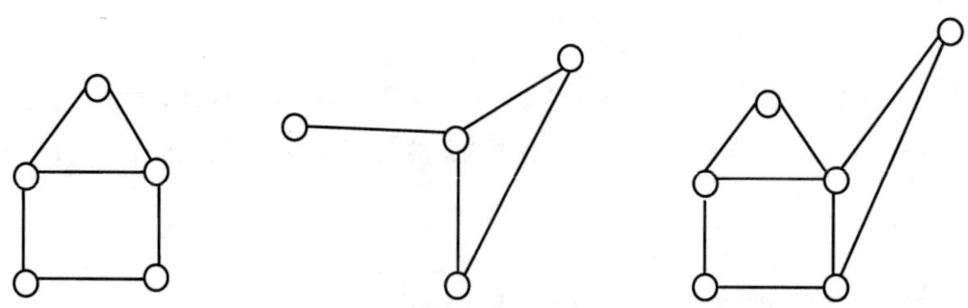

(a) G_1 **(b)** G_2 **(c)** $G_1 \cup G_2$

Figure 1.38

For two vertex disjoint graphs G_1 and G_2, $G_1 \cup G_2$ is denoted by $G_1 + G_2$ and is called the sum of graphs G_1 and G_2.

The ***join*** of G_1 and G_2 is the super graph of $G_1 + G_2$ in which each vertex of G_1 is adjacent to every vertex of G_2. That is, the join of G_1 and G_2 consists of $G_1 \cup G_2$ and all edges joining G_1 with G_2. For instance, in Figure 1.39 the ***join*** $G_1 + G_2$ consists of $G_1 \cup G_2$ and all lines joining V_1 with V_2.

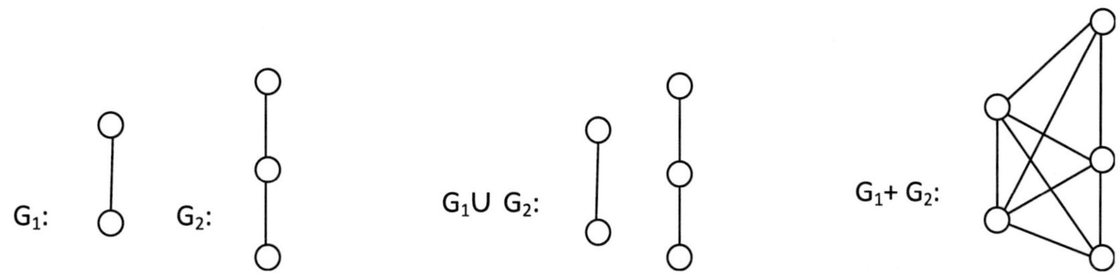

Figure 1.39 The union and join of two graphs

Note 2: The finite union of graphs is defined by means of associativity; in particular, if G_1, G_2, G_3, ………, G_r are pair wise vertex disjoint graphs, each of which is isomorphic to G, then $G_1 + G_2 + G_3 + \ldots\ldots\ldots + G_r$ is denoted by r G.

Intersection: Given two graphs G_1 and G_2 with at least one vertex in common then their intersection will be a graph such that

$$V(G_1 \cap G_2) = V(G_1) \cap V(G_2) \text{ and}$$

$$E(G_1 \cap G_2) = E(G_1) \cap E(G_2)$$

The intersection of graphs G_1 and G_2 shown in Figure 1.38 (a and b) is shown in Figure 1.40.

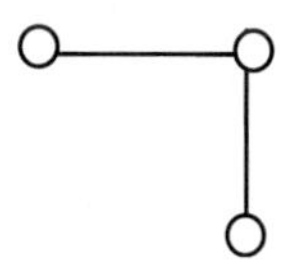

Figure 1.40 $G_1 \cap G_2$

Complement: The complement G^c of G is defined as a simple graph with the same vertex set as G and where two vertices u and v are adjacent only when they are not adjacent in G (Figure 1. 41).

Figure 1.41 Graphs G_1 and G_2 with its complements G_1^c and G_2^C

A simple graph G is called self-complementary if $G \cong G^c$ (Figure 1.42). Note that, G^c is also a simple graph and that $(G^c)^c = G$. Also If $|V(G| = n$, then clearly, $|E(G) + |E(G^c)| = n(n-1)/2$. A self complementary graph is isomorphic with its complements. Of course, this can only occur if G and its complement have the same size. Not only are the orders the same and sizes the same of two isomorphic graphs, so too are the degrees of their vertices.

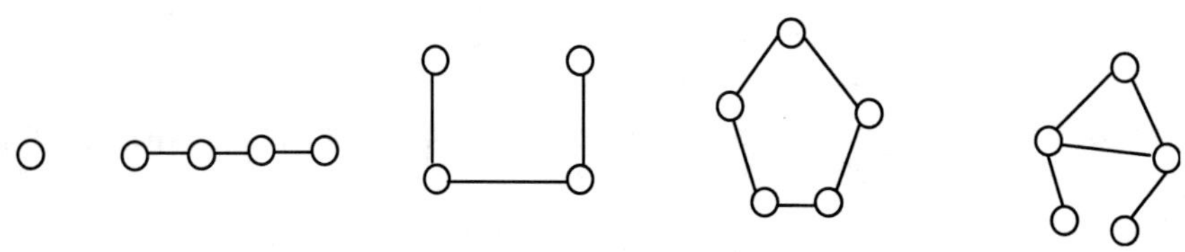

Figure 1.42 Some Self complementary graphs

Product of two graphs: To define the ***Cartesian product*** $G_1 \times G_2$ of two graphs (not necessarily vertex disjoint) consider any two points $u = (u_1, u_2)$ and $v = (v_1, v_2)$ in $V_1 \times V_2$. Then u and v are adjacent in $G_1 \times G_2$ whenever $[u_1 = v_1$ and u_2 adj $v_2]$ or $[u_2 = v_2$ and u_1 adj $v_1]$. That is, the Cartesian product $G_1 \times G_2$ has vertex set $V(G_1 \times G_2) = V(G_1) \times V(G_2)$. In general, every vertex of $G_1 \times G_2$ is an ordered pair (u, v) where $u \in V(G_1)$ and $v \in V(G_2)$. Two distinct vertices (u, v) and (x, y) are adjacent in $G_1 \times G_2$ if either u = x and $vy \in V(G_1)$ or v = y and $ux \in E(G_1)$. Figure 1.43 shows the Cartesian product of graphs G_1 and G_2 .

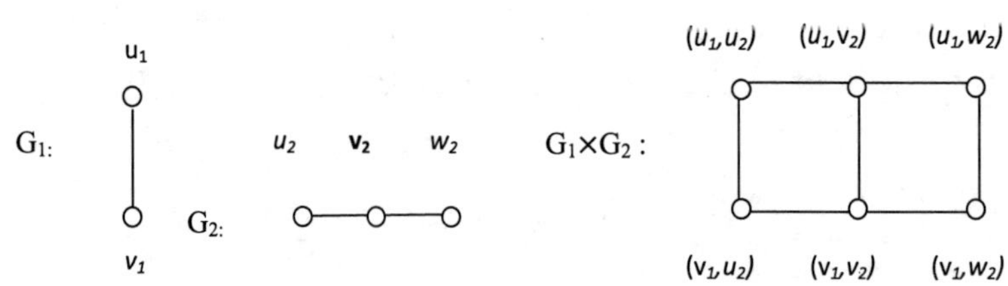

Figure 1.43 The Cartesian product of two graphs

Some additional comments about Cartesian productof graphs are worth mentioning. First the definition of Cartesian product tells us that the order in which the graphs G_1 and G_2 is written is structurally irrelevant. (i.e., $G_1 \times G_2$ and $G_2 \times G_1$ are the same graph). Obviously, they are isomorphic graphs.

There is an informal way of drawing the graph $G_1 \times G_2$ or $G_2 \times G_1$ that does not require the labeling of vertices. Replace each vertex u of $G_1$by a copy of G_2 (u) of the graph G_2. Then, If the vertices u_1 and u_2 in G_1 are adjacent, join corresponding vertices of $G_2(u_1)$ and G_2 (u_2) by an edge. If u_1 and u_2 are not adjacent in G_1, then we add no edges between the $G_2(u_1)$ and $G_2(u_2)$ (Figure 1.44).

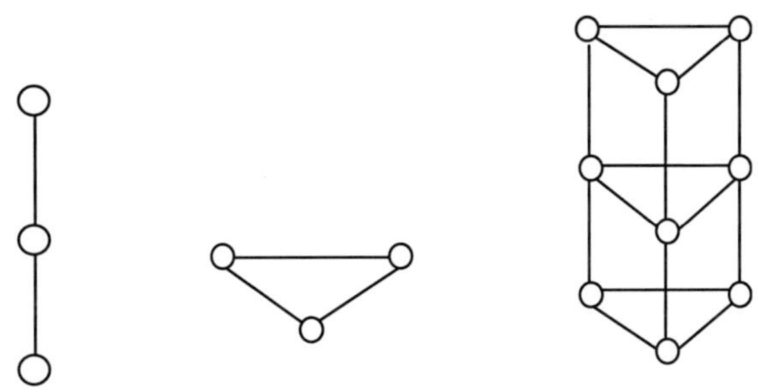

Figure 1.44 Cartesian product of two graphs P_3 and K_3

An especially important class of graphs known as cubes are most naturally expressed in terms of products. The n –cube Q_n (or hyper cubes) is defined recursively by $Q_1 = K_2$ and $Q_n = K_2 \times Q_{n-1}$. The n-cube can also be defined as the graph whose vertex set is the set of ordered n- tuples of 0's and 1's and where two vertices are adjacent if their ordered n-tuples differ in exactly one position (coordinate). Thus Q_n has 2^n points which may be labeled $a_1, a_2 \ldots\ldots a_n$, where each a_i is either 0 or 1. Two points of Q_n are adjacent if their binary representation differ at exactly one place. The n-cubes for n = 1, 2, 3 are shown in Figure 1.45, where the vertices are appropriately labeled by n-bit strings.

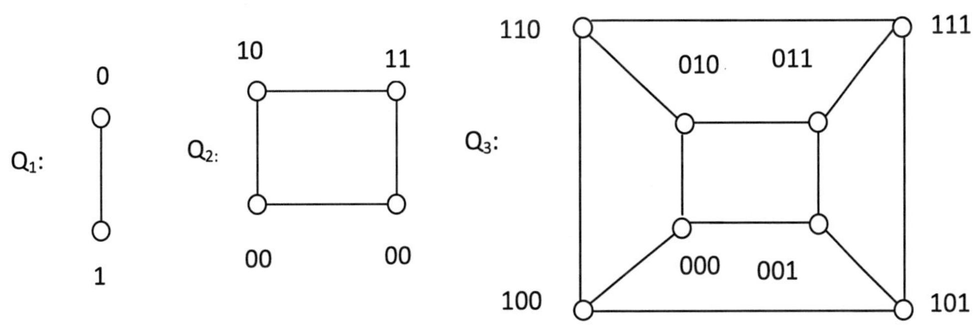

Figure 1.45 The n cubes for n = 1, 2, 3

Composition or Lexicographic Product: The composition $G = G_1[G_2]$ also has $V = V_1 \neq V_2$ as its point set and $u = (u_1, u_2)$ is adjacent with $v = (v_1, v_2)$ whenever $[u_1 \text{ adj } v_1]$ or $[u_1 = v_1$ and u_2 adj $v_2]$. For the graphs G_1 and G_2 of Figure 1.43 both compositions $G_1[G_2]$ and $G_2[G_1]$ which are obviously not isomorphic, are shown in Figure 1.46.

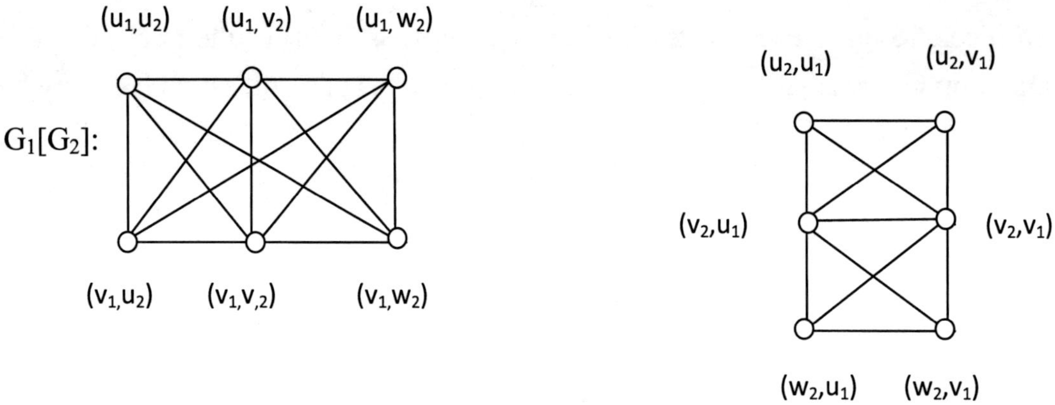

Figure 1.46 Two compositions of graphs

***Normal product or strong product*:** The normal product $G_1 \circ G_2$ of two simple graphs G_1 and G_2 is the graph with $V(G_1 \circ G_2) = V_1 \times V_2$ wherein (u_1, u_2) and (v_1, v_2) are adjacent in $G_1 \circ G_2$ if, and only if, either

(i) $u_1 = v_1$ and u_2 adjacent to v_2, or

(ii) u_1 is adjacent to v_1 and $u_2 = v_2$, or

(iii) u_1 is adjacent to v_1 and u_2 is adjacent to v_2.

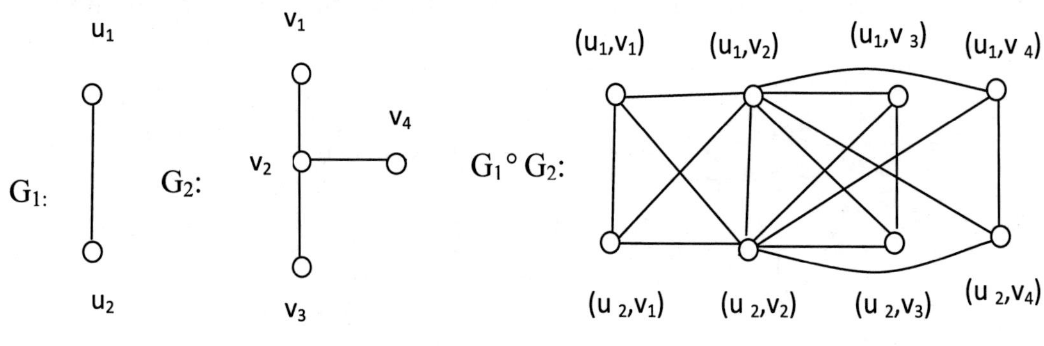

Figure 1.47

For example, Figure 1.47 (c) is the normal product of graphs G_1 and G_2 of Figure 1.47 (a) & (b).

***Tensor or Kronecker product*:** The Tensor product $G_1 \oplus G_2$ of two simple graphs G_1 and G_2 is the graph with $V(G_1 \oplus G_2) = V_1 \times V_2$ and where in (u_1, u_2) and (v_1, v_2) are adjacent in $G_1 \oplus G_2$, if and only if, u_1 is adjacent to v_1 in G_1 and u_2 is adjacent to v_2 in G_2

We note that $G_1 \circ G_2 = (G_1 \times G_2) \cup (G_1 \oplus G_2)$. For example for the graph of

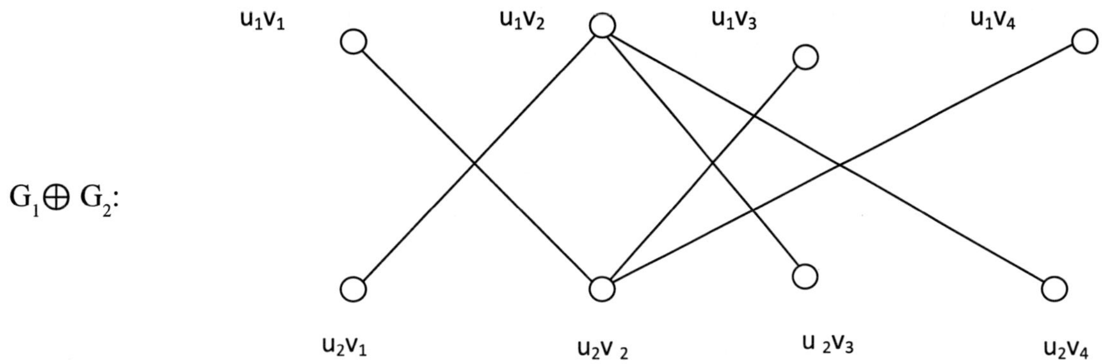

Figure 1.48 $G_1 \oplus G_2$

G_1 and G_2 of Figure 1.47 (a) & (b) the tensor product $G_1 \oplus G_2$, is the one shown in Figure 1.48.

1.6 SUB GRAPHS

If G and H are two graphs with vertex sets V(H), V(G) and edge sets E(H) and E(G) respectively such that $V(H) \leq V(G)$ and $E(H) \leq E(G)$ then we call H asa ***sub graph*** of G (or G as a super graph of H). We also say that G ***contains*** H as a sub graph. (Figure 1.49). More generally, A sub graph of G is a graph having all its points and lines in G. Obviously, the graph G and null graph are sub graphs of G. This, in fact, is the definition if the graphs H and G are labeled (that is, if the vertex sets of H are G have been specified). For unlabeled graphs H and G, we say that H is a sub graph of G if for any labeling of the vertices of H and G, the graph H is isomorphic to a sub graph of G. Consequently, for the graphs H and G of Figure 1.49, H is a sub graph of G.

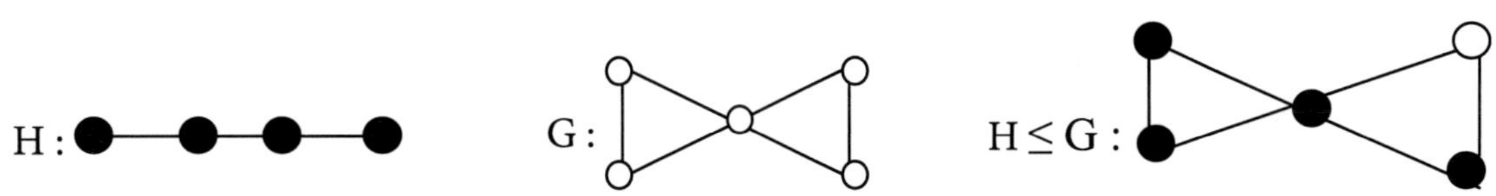

Figure 1.49 A sub graph H of a graph G

A digraph H is called a ***sub digraph*** of a digraph D, if $V(H) \leq V(D)$ and $E(H) \leq E(D)$. For the graph of Figure 1.10, an epidemic in a herd can be described by dividing the herd into classes (sub graphs) consisting of susceptible, exposed, infectious, recovered and dead ones. As another example, we take a simple food web consisting of cultivated plant (central element), weeds, different type of pests, and their predators, in which, the cultivated plant as a sub graph can consists of elements representing the relative water holding capacity, the evapotranspiration of the plant, the biomass growth etc. The soil element as a sub graph can consist of water content of soil, different nutrient content, temperature of different layers of soil, the water run off in the soil, the evaporation or water holding capacity of soil. A sub graph H of a given directed graph G is called a **spanning sub graph** of G if H contains all vertices of G. The directed graph shown in

Figure 1.50(b) is a sub graph of the graph shown in Figure 1.50(a). Directed graph in Figure 1.50(c) and 1.50(d) are two of the $2^4 = 16$ possible spanning sub graphs.

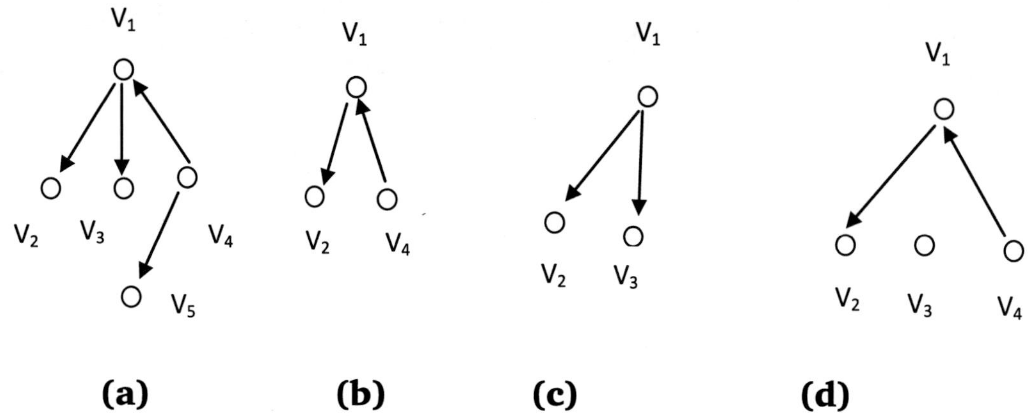

Figure 1.50 Sub graphs

A sub graph H of a graph G is a ***proper sub graph*** of G if either V(H) .≠ V(G) or E(H) . ≠ E(G), and if a sub graph of G has the same vertex sets as G (i.e., if V(H) = V(G)), then we say that H is a ***spanning sub graph*** of G. If G is a graph with vertex set V and U is the subset of V then the ***sub graph G(U) of G induced by U*** is defined to be the graph having vertex set U and edge set consisting of those edges of G that have both ends in U (i.e, A sub graph F of a graph G is called ***induced sub graph*** of G if whenever u and v are vertices of F and uv is an edge of G, then uv is an edge of F as well). Similarly if F is a non-empty subset of the edge set E of G then the sub graph G(F) of G induced by F is the graph whose vertex set is the set of ends of edges in F and whose edge set is F. If from a graph G with vertex set V, we delete a subset U of V and all edges which have a vertex in U as an end point then (G-U) is called a ***vertex deleted sub graph*** and similarly if a subset F of E is deleted from G then (G-F) denotes the sub graph of G with Vertex sets V and edge set E-F, then (G-F) is called ***edge deleted sub graph*****(Figure 1.51).**

Note that when a vertex is deleted from G, all the edges incident to it are also deleted from G, whereas the deletion of an edge from G does not affect the vertices of G. That is, removal of an edge from G leaves its end points in place, whereas the removal of a vertex necessitates the removal of edges that vertex associated as an end point. By deleting from a graph G all loops and in each collection of parallel edges all edges but one we obtain a simple spanning sub graph of G, called the underlying simple graph of G.

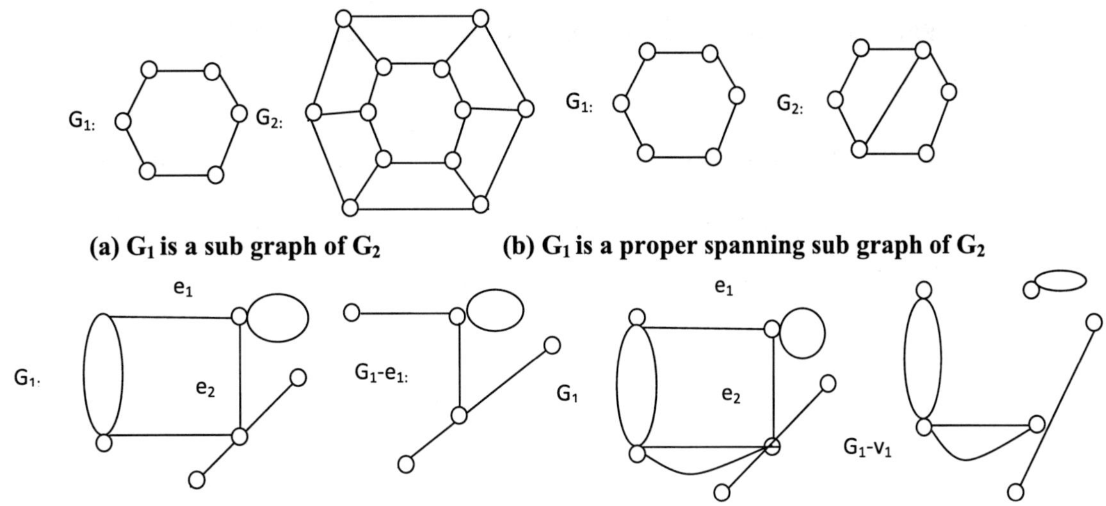

(a) G_1 is a sub graph of G_2 **(b) G_1 is a proper spanning sub graph of G_2**

(c) G_1-e_1: A case of edge deleted sub graph (d) G_1-v_1:A case of vertex deleted sub graph

Figure 1.51 Showing various sub graphs

The complement of a sub graph H (V', E') of a graph G(V, E) is the sub graph H^c (V, E-E'). That is, complement of a sub graph H is obtained by removing edges of H from G (Figure 1.52).

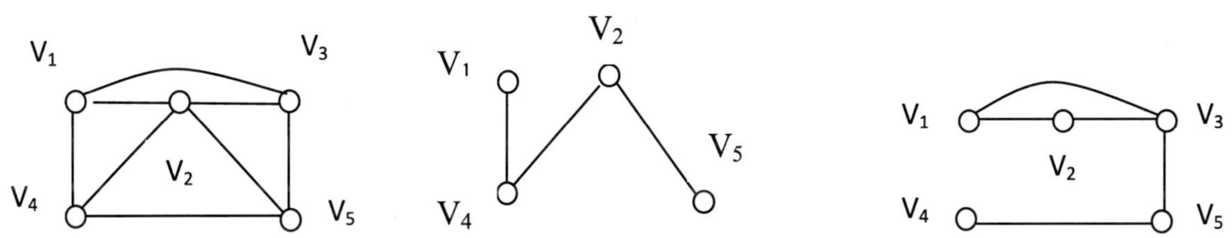

(a) Graph G (b) Sub graph H (c) Complement of sub graph H

Figure 1.52

Note 2:

(i) Every graph is its own sub graph.

(ii) A sub graph of a sub graph of G is a sub graph of G.

(iii) A single vertex in a graph G is a sub graph of G.

(iv) A single edge in G, together with its end vertices, is also asub graph of G.

1.7 SOME APPLICATIONS OF GRAPHS

At this stage, we will only state some of the problems in which graphs can be applied. The solution of these problems is left presently. When you move to later units you will find a detailed discussion of their solution. From among the hundreds of applicationswe cite below three examples that are of great significance.

1.7.1 The MTNL's Networking Problem

Suppose Mahanagr Telephone Nigam Limited (MTNL) is interested in identifying those lines that must stay in service to avoid disconnecting the network given in Figure 1.53.

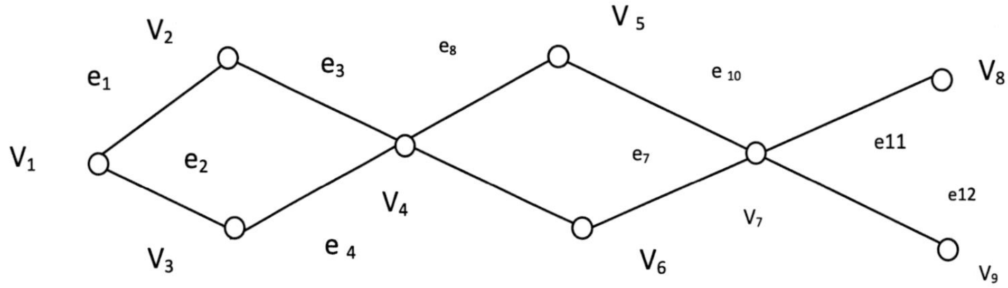

Figure 1.53 Network of Telephone Lines

This is a typical problem of graph theory which you will read in further units.

1.7.2 Travelling Salesman Problem

A salesman has to visit several cities. He is required to visit each city exactly once. Graph theory can be used to solve this transportation system.

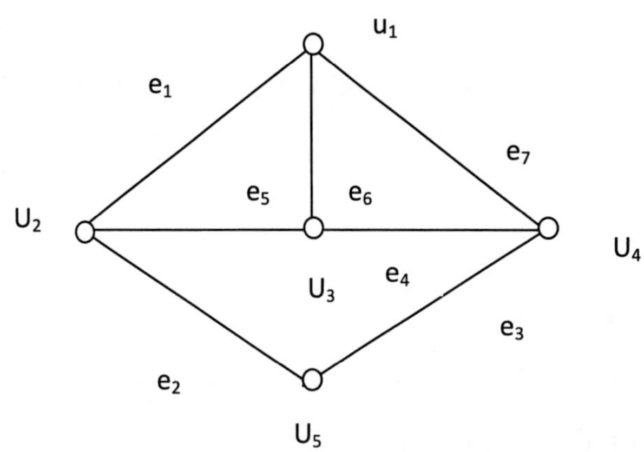

Figure 1.54 Travelling Salesman's Territory

Here u_1.......... u_5 represents cities and e_1....... e_7 the routes connecting them.

1.7.3 Satellite Channel Problem

Suppose six TV companies apply for frequency allotment. If the relay centers of these companies are not less than 1000 km then only the same frequency can be allotted without interference. The aim is to assign as small a number of different frequency as possible.

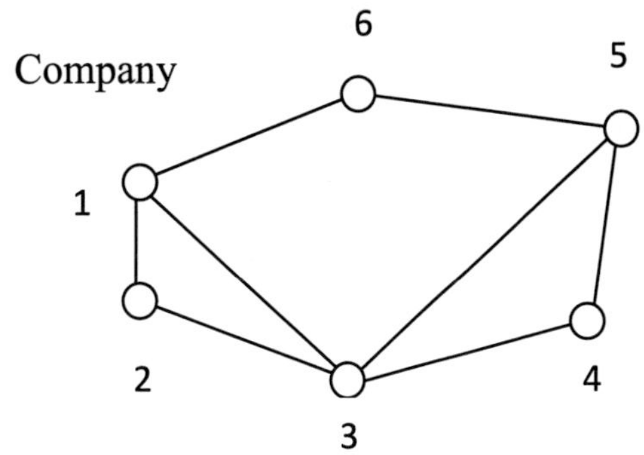

Figure 1.55 Relay centers and their distance

The line joining two companies shown that they have relay centers less than 1000 km apart. This is a typical problem of graph theory. Here as a obvious case, we can assign same frequency to company 1 and 4, 2 and 5 and 3 and 6 respectively (Figure 1.55).

1.7.4 Crop Pest and Disease Spread Control Problem in Agricultural Landscapes

Supposethe landscape network of a country, isrepresented as a set of nodes (nodes positioned at the geographical centroid of each state in the country) linked by edges thenthe pest or pathogen movementin the network isindicated by adjacencies among states. More over, we assume that spread of pathogen or pest is facilitated by host density andthe weighted mean proportion of crop acreage along the length of the edge, provides a measure of the host availability across two states in the country. Removal of an edge above a particular landscape resistance to transmission (LRT) threshold determines which states are connected when disease or pest spread is likely (resulting from low host density), with low threshold LRT indicating which all states stay connected (resulting from high host density). By characterizing the scales at which regional connectivity become disrupted, better strategies for dealing with invasive pathogens, pests and strategies for changing cropping patterns can be formulated in the graph theoretic framework. Information about connectivity is also important input for evaluations of pest or pathogen risk at any given location, since more highly connected locations experience higher risk.

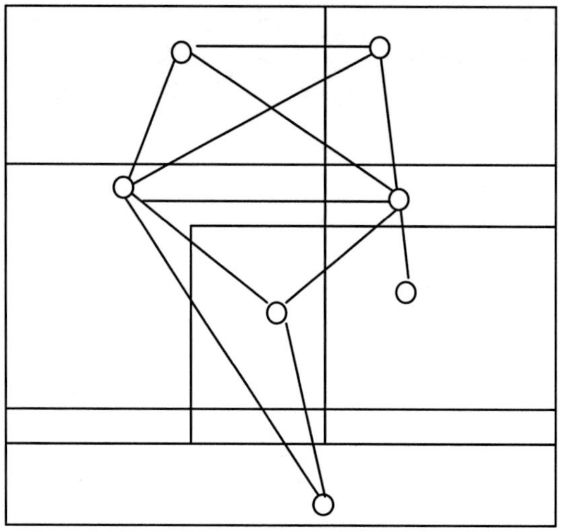

Figure 1.56 A graph suited to a situation where Landscape Matrix is divided by Geographical boundaries

SUMMARY

1. Graph Theory is the theory that is referring to a set of **vertices** connected by **edges** (or lines) and is actually the study of mathematical graphs (mathematical structures used to model pair wise relations between objects from a certain collection). The set of vertices is usually denoted by V(G) and the set of edges by E(G).

2. A graph may be **undirected**, meaning that there exists no distinction between the two vertices or it may be **directed** (i.e., its edges may be directed from one vertex to another). If a vertex is joined by more than one edge (or there is another edge with the same end vertices), then it is called a **multigraph**, otherwise it is simple. A graph without any edge is called a **null** graph.

3. A graph is called a **simple** graph if it is undirected and has no loops and no more than one edge between any two different vertices. In a simple graph each edge is a pair of distinct vertices. The edge which starts and ends in the same vertex is called a **loop**. An edge connects two vertices. An edge is **parallel**/multiple if there is another edge with the same end vertices. These two vertices are said to be **incident** to that edge or equivalently that edge incident to those two vertices.

4. An edge, uv in a directed or undirected graph G is said to be an edge atthe vertices u and v, and the two vertices are said to be **adjacent** to each other. In this case, we also say that u and v are **neighbours**. For an undirected graph, G and a vertex, u Є V(G), the set of all neighbours of u is denoted N(u) and is given by N(u) = {v Є V(G) :uv Є E(G)}. A ***regular graph*** is a graph where each vertex has the same number of neighbours. A D regular digraph is a graph where, od(v) = id (v), for every vertex v of D.

5. The **degree** of a vertex v in a graph G is the number of edges attached to that particular vertex, with loops being counted twice. In a directed graph *G*, the **in-degree**, $deg_{in}(u)$ (**out-degree,**

deg_{out} (u)) of a vertex u is given by the number of edges that terminate (start) at *u*. If the set of edges E(G) is finite, then the total sum of vertex degrees is equal to twice the number of edges. That is, the total degree of a graph is equal to two times the number of edges (loops included). A sequence formed by the degrees of vertices of G is called a ***degree sequence*** of G. The degree sequence $d = (d_1, d_2, \ldots d_n)$ is **graphic** if there is a simple undirected graph with degree sequence d.

6. A graph G is a **bipartite**, if its vertex set V can be partitioned into two non-empty subsets X and Y in such a way that every edge of G joins X with Y. If every vertex in X is joined to every vertex of Y, then we call it a **complete bipartite graph** denoted by $K_{m, n}$, where m and n represent vertices in X and Y respectively.

7. Graphs in which edges either exist or do not exist, and in which all edges have the same significance are called **unweighted** graphs. when weights are assigned to each of the edges the corresponding graph is called a **weighted** graph.

8. A **complete graph** has an arc connecting each pair of vertices.

9. For graphs G with the set of vertices V(G) and the set of edges E(G) and G'with the set of vertices V(G') and the set of edges E(G'), G' is a ***subgraph*** of a graph G if V (G') is the subset of V(G) and E(G') is the subset of E(G).

10. A sub graph F of a graph G is called an ***induced subgraph*** of G if whenever u and v are vertices of F and uv is an edge of G, then uv is an edge of F as well . A vertex deleted sub graph of a graph G is a sub graph G-v, obtained by deleting from G the vertex v and all edges incident to it. Similarly, an edge deleted sub graph Of G is a sub graph G-e, obtained by deleting the edge e, A subgraph G' is a **spanning sub graph** of a graph G if it has the same vertex set as the graph G.

EXCERCISES

1. Draw the diagram of the following graphs G(V, E):

   (a) $V = \{v_1, v_2, v_3, v_4\}$, $E = \{(v_1, v_2), (v_4, v_1), (v_3, v_1), (v_3, v_4)\}$

   (b) $V = \{v_1, v_2, v_3, v_4, v_5, v_6\}$, $E = \{(v_1, v_4), (v_1, v_6), (v_4, v_6), (v_{3,} v_2), (v,_3 v_5), (v_2, v_5)\}$

   (c) $V = \{v_1, v_2, v_3, v_4, v_5\}$, $E = (v_1, v_2), (v_1, v_2), (v_1, v_4), (v_1, v_5), (v_2, v_4), (v_3, v_4), (v_3, v_4), (v_3, v), (v_3, v_5)\}$.

2. Find the number of vertices, number of edges and degree of each vertex in the given undirected graph. Identify all isolated and pendant vertices?

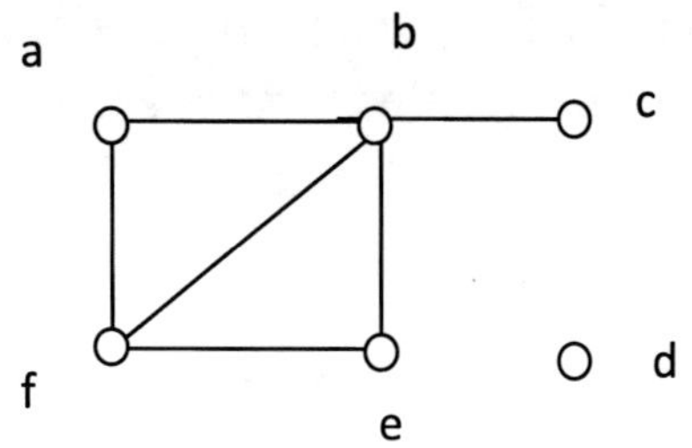

(a)

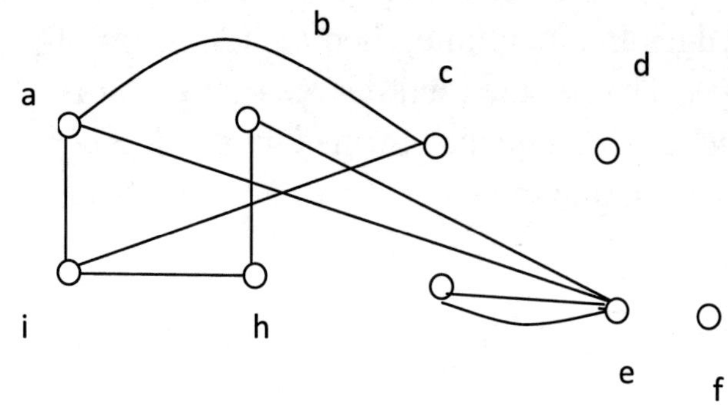

(b)

3. Find the number of vertices, no of edges and in degree and out degree of each vertex in the given directed graph?

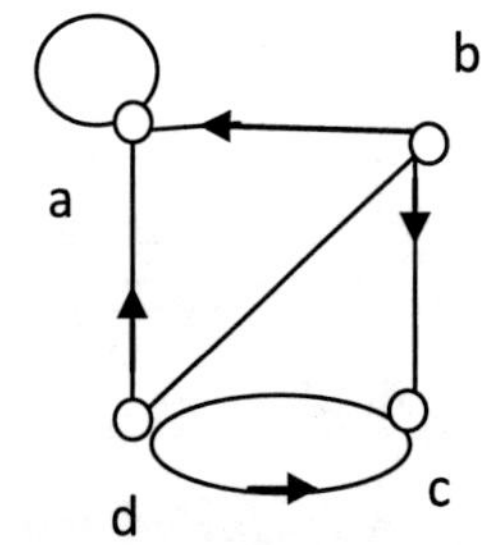

4. For the following graph G shown in figure below, find (i) degree of v_5 and v_1 (ii) the vertices incident with e_5 (iii) Neighbours of v_3?

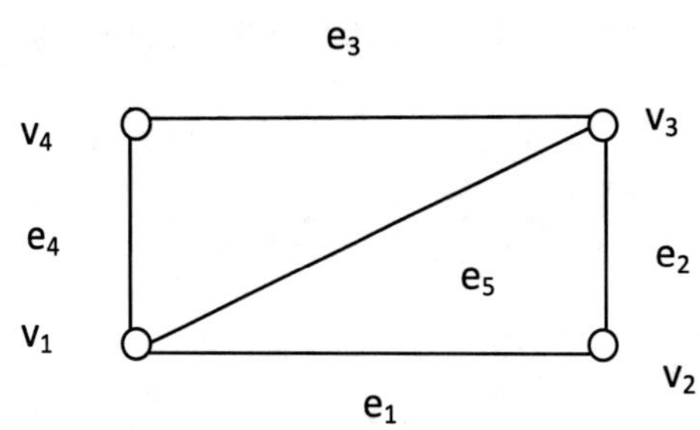

5. Draw a graph with five vertices v_1, v, v_3, v_4, v_5 such that deg $(v_1) = 3$, deg(v_2) is an odd vertex, deg$(v_3) = 2$ and vertices v_4 and v_5 are adjacent?

6. Give an example of a graph such that every vertex is adjacent to two vertices and every edge is adjacent to two edges?

7. For each of the following sequences, [a] 4, 4, 4, 3, 2. [b] 5, 5, 4, 3, 2, 1, determine if there exists a graph whose degree sequence is the one specified. In each case, either draw a graph or explain why no such graph exists?

8. Is the degree sequence (a) 2, 3, 3, 4, 5, 6, 7 (b) 1, 3, 3, 3 graphical?

9. Draw 3-regular graphs with (i) six and (ii) eight vertices.

10. Determine whether the graphs in Figure (a-e) is bipartite. In each case, either draw the bipartition sets or explain why it is not bipartite?

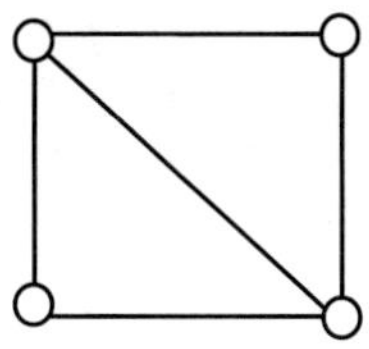

(a)

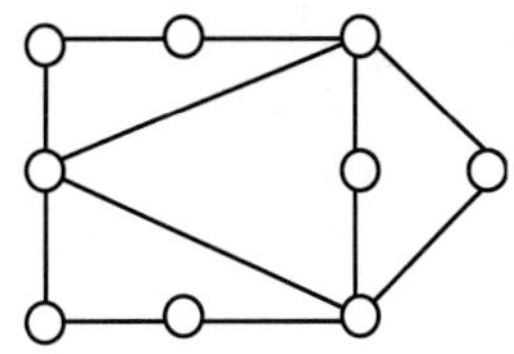

(b)

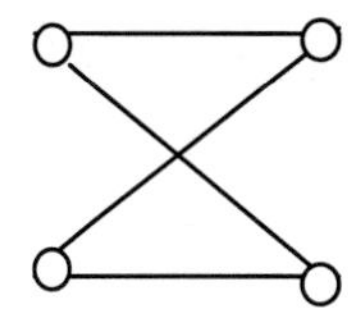

(c)

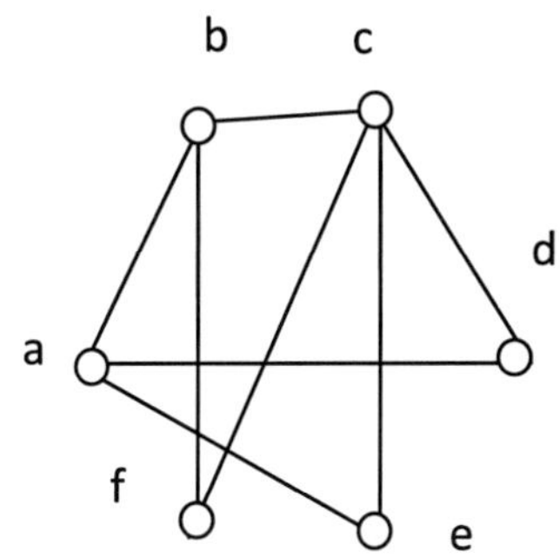

(d)

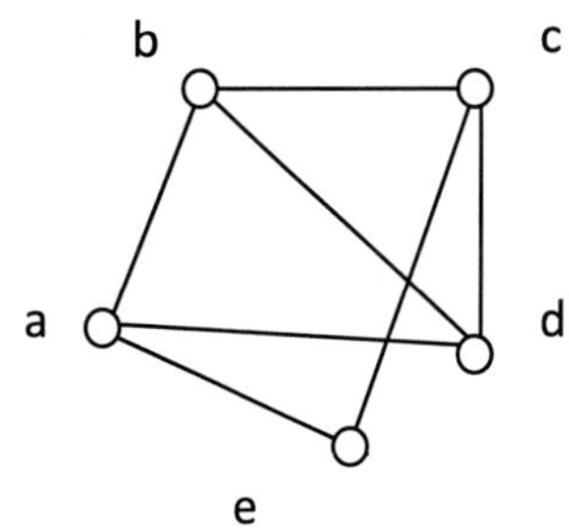

(e)

11. For the following graphs G_1 and G_2 of graphs, draw $G_1 + G_2$ and $G_1 \times G_2$.

(i) $G_1 = k_5$ and $G_2 = K_2$ (ii) $G_1 = C_5$ and $G_2 = K_1$

12. For the graphs G and G' of figure below find the their intersection and union?

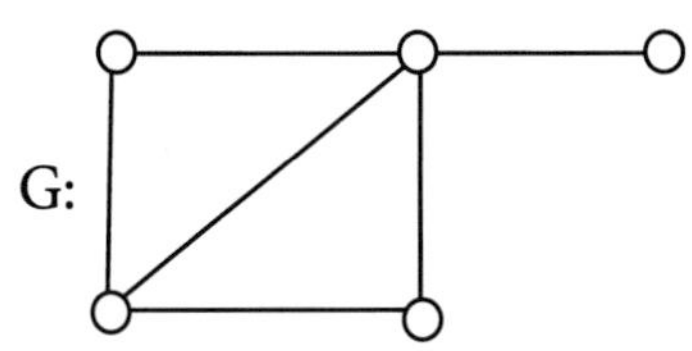

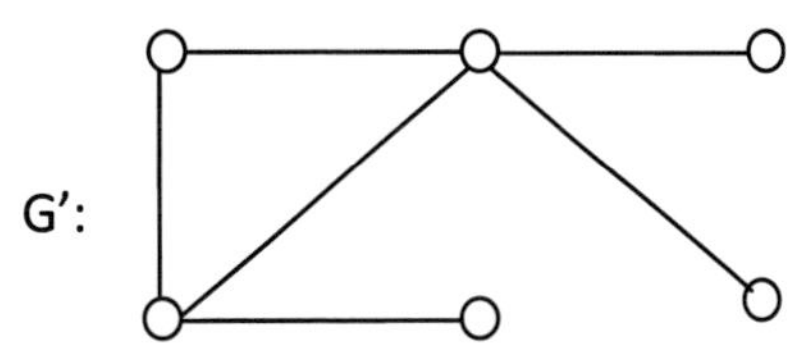

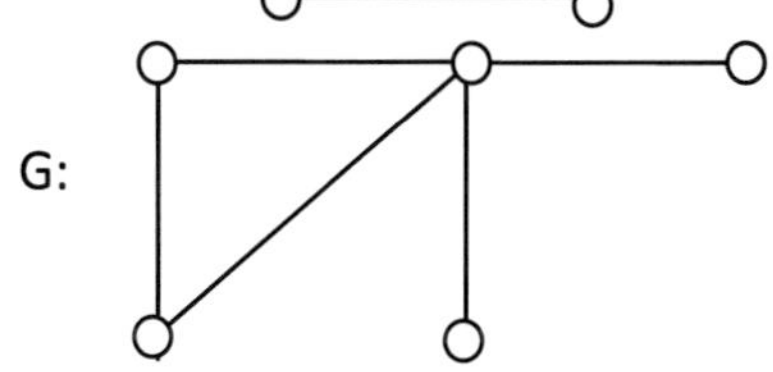

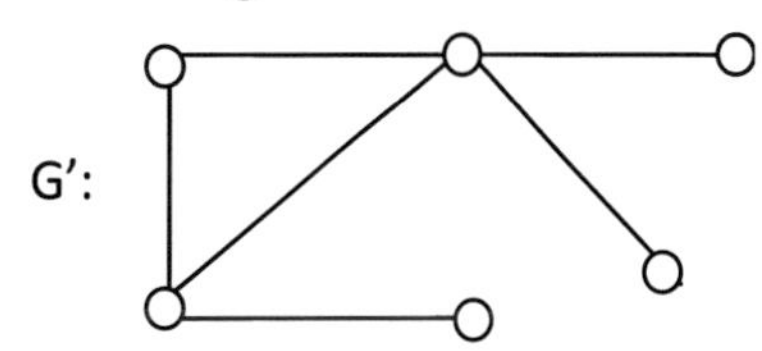

13. Find the complement of each of the graphs shown below?

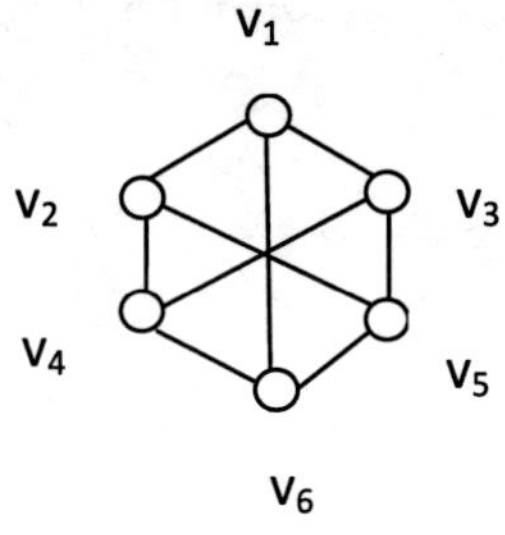

(a)

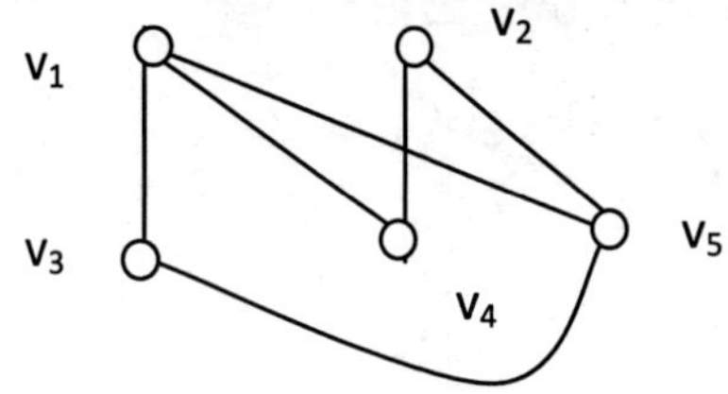

(b)

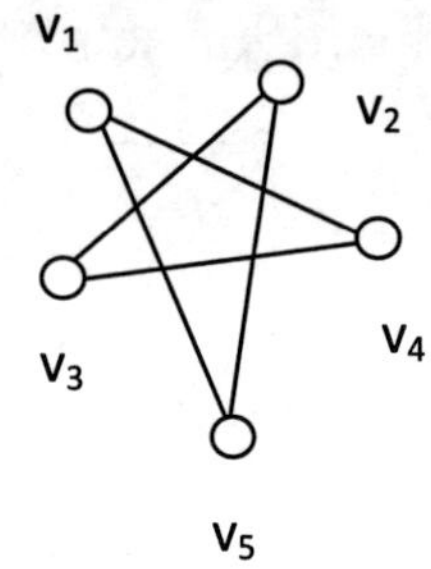

(c)

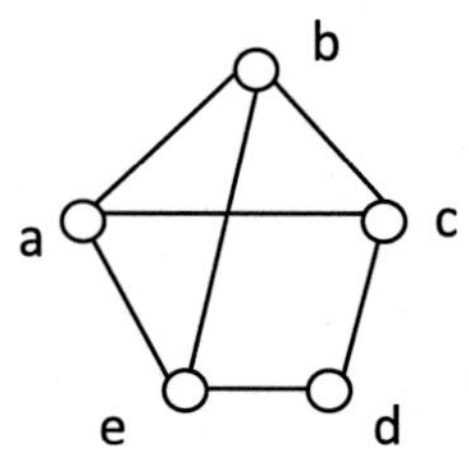

(d)

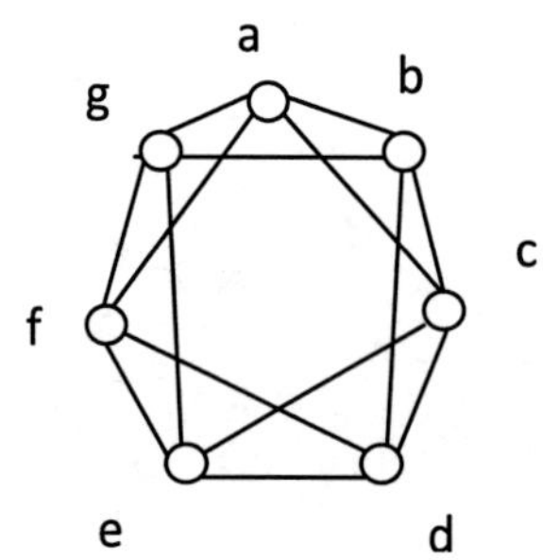

(e)

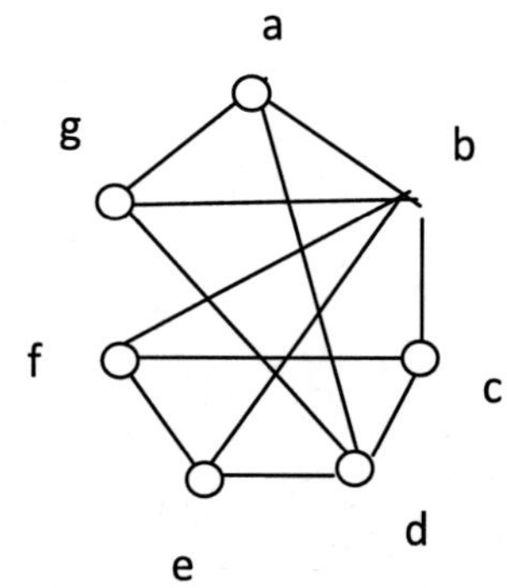

(f)

2

MATRIX REPRESENTATION OF GRAPHS

2.0 INTRODUCTION

Although it is more convenient to represent a graph by a diagram of points connected by lines, such a structure may be unsuitable if one wish to store a large graph in a computer. One of the ways of storing a simple graph is by listing adjacent vertices of the graph. That is, A graph is completely determined by either its adjacencies or its incidences. Other useful representations involve matrices. Indeed, with a given graph, adequately labeled, there are associated several matrices including the adjacency matrix, incidence matrix, and cycle matrix. It is often possible to make use of these matrices in order to identify certain properties of a graph. We present here two types of matrices commonly used to represent graphs. One is based on adjacency of vertices and the other on incidence of vertices and edges. Adjacency and incidence matrices are dependent on how vertices and edges of G are labeled. In any case, the adjacency matrix is a n × n matrix where every entry on the main diagonal is zero. The number of 1's in row i (or column i) is the degree of the vertex i, while the number of 1's in row i of the incidence matrix is also the degree of $v_{i.}$ The number of 1's in each of its columns is 2 since there are exactly two vertices incident with every edge.

2.1 ADJACENCY MATRIX

Let G be a graph with n vertices v_1, v_2, ... v_n. The ***adjacency matrix*** (or Connection matrix) of G with respect to this particular listing of n vertices is the n × n matrix with 1 as its $(i, j)^{th}$ entry when i and j are adjacent, and 0 as its $(i, j)^{th}$ entry when they are not adjacent. In other words, if its adjacency matrix is $A(G) = (a_{ij})$, then,

$$aij = \begin{cases} 1 & \text{if } v_i \text{ is adjacent to } v_j \text{ (i.e., } v_i v_j \in E(G)) \\ 0 & \text{othertwise} \end{cases}$$

It is denoted by A(G) or A_G. If G has no loops, then all the entries to the main diagonal will be zero and if G has no parallel edges then the entries of A(G) are either zero or one. Given a n × n symmetric matrix A = (a_{ij}), in which all the entries are non-negative integers, one can associate with it a graph G whose adjacency matrix is A by letting G have n vertices and joining vertex i to vertex j by a_{ji} edges. In general, because of the correspondence between graphs and matrices, any graph theoretic concept is reflected in the adjacency matrix.

The following is a adjacency 5×5 matrix representation for the simple undirected graphs of Figure 2.1.

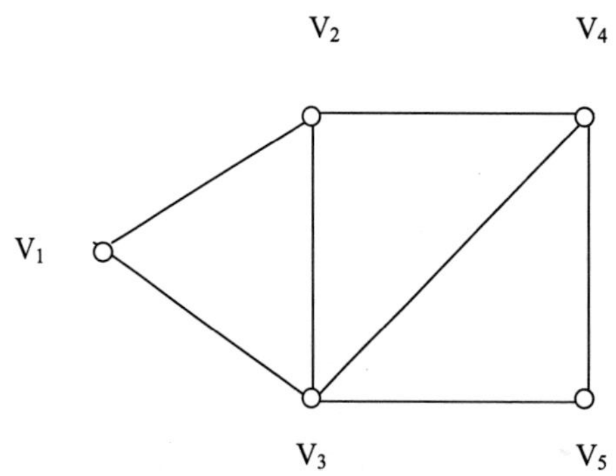

Figure 2.1

$$A = \begin{array}{c} \\ v_1 \\ v_2 \\ v_3 \\ v_4 \\ v_5 \end{array} \begin{array}{c} \begin{array}{ccccc} v_1 & v_2 & v_3 & v_4 & v_5 \end{array} \\ \begin{pmatrix} 0 & 1 & 1 & 0 & 0 \\ 1 & 0 & 1 & 1 & 0 \\ 1 & 1 & 0 & 1 & 1 \\ 0 & 1 & 1 & 0 & 1 \\ 0 & 0 & 1 & 1 & 0 \end{pmatrix} \end{array}$$

For a Directed graph G with n vertices, n × n adjacency matrix is defined as

$$A_{ij} = \left.\begin{matrix} 1 \\ 0 \end{matrix}\right\} \begin{matrix} \text{if edges beginning at } v_i \text{ and ending at } v_j \\ \text{otherwise} \end{matrix}$$

The non zero elements in the matrix is equal to the number of edges in the digraph. The adjacency matrix for a digraph does not have to be symmetric, because there must not be an

Edge from a_j to a_i when there is an edge from a_i to a_j. The following is a adjacency 5 × 5 matrix representation for the simple undirected graphs of Figure 2.2.

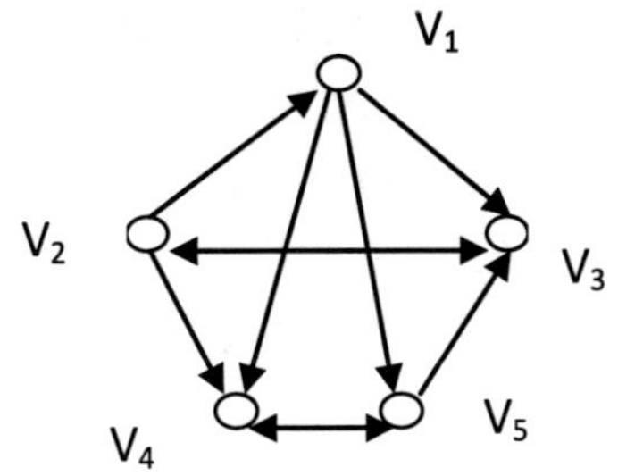

$$A = \begin{array}{c} \\ v_1 \\ v_2 \\ v_3 \\ v_4 \\ v_5 \end{array} \begin{array}{c} \begin{array}{ccccc} v_1 & v_2 & v_3 & v_4 & v_5 \end{array} \\ \begin{pmatrix} 0 & 0 & 1 & 1 & 1 \\ 1 & 0 & 0 & 1 & 0 \\ 0 & 1 & 0 & 0 & 0 \\ 0 & 0 & 0 & 1 & 1 \\ 0 & 0 & 1 & 0 & 0 \end{pmatrix} \end{array}$$

Figure 2.2 A Digraph

When graph G is a multigraphs, then a_{ij} denote the number of edges (v_i, v_j). Thus, only adjacency matrix is used to represent multigraphs. For a multigraphs G consists of n vertices, an $n \times n$ adjacency matrix $A = [a_{ij}]$ is defined as

$$A_{ij} = \left.\begin{matrix} m \\ 0 \end{matrix}\right\} \begin{matrix} \text{if there are one or more edges between vertices } v_i \text{ and } v_j; m = \text{the number of edges} \\ \text{otherwise} \end{matrix}$$

The following is the adjacency matrix representation for a multigraph with four vertices of the Figure 2.3.

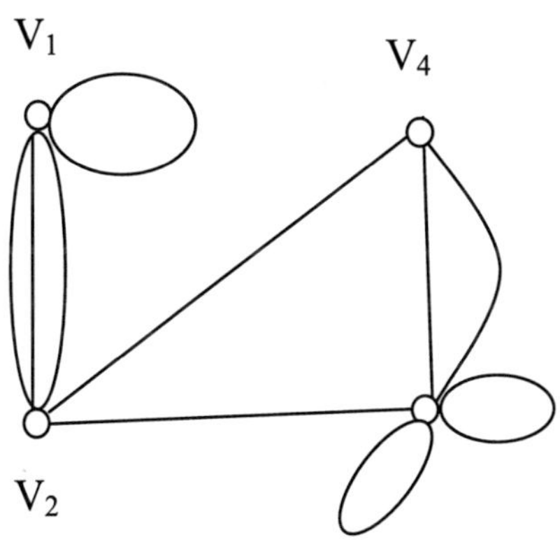

$$A = \begin{array}{c} \\ v_1 \\ v_2 \\ v_3 \\ v_4 \end{array} \begin{array}{c} \begin{array}{cccc} v_1 & v_2 & v_3 & v_4 \end{array} \\ \begin{pmatrix} 1 & 3 & 0 & 0 \\ 3 & 0 & 1 & 1 \\ 0 & 1 & 2 & 2 \\ 0 & 1 & 2 & 0 \end{pmatrix} \end{array}$$

Figure 2.3 A Multigraph

A number of results about the adjacency matrix are as follows:

(a) For Undirected simple graph

1. The adjacency matrix completely defines a simple graph. if the graph G is simple, the degree of the vertex v_i equals the number of 1's in the ith row (j^{th} Column) of A(G).
2. The adjacency matrix is symmetric.
3. Any element of the adjacency matrix is either 0 or 1 .
4. The ith row in the adjacency matrix is determined by the edges which originate in the node v_i.
5. G is null $\Leftrightarrow$ A(G) is the zero matrix of order n.

6. The adjacency matrix depends upon the ordering chosen for the vertices. For different orderings of elements of V, we get different (n!) adjaceancy matrices of the same graph. But any one of the adjacency matrices of G can be obtained from another adjacency matrix of the same graph G by interchanging some of the rows and corresponding columns of the matrix. Hence we can take any adjacency matrix of the graph G.

7. The adjacency matrix has n^2 elements. They are useful for determination of paths in graphs. For example, it is easy to show that there is a path of length (i.e., having r edges) from v_i to v_j if and only if $A^r = 1$, where A^r is the rth matrix product.

8. A graph G is connected if and only if the matrix

$$B = A + A^2 + \ldots\ldots + A^{n-1}$$

Has no zero entries off the main diagonal, where A is the adjacency matrix of G with n vertices of the graph.

(b) Directed graphs

1. The number of elements in the ith row whose value is 1 is equal to the out degree of the vertex v_i.

2. The adjacency matrix for a digraph may or may not be symmetric matrix, since there may not be an edge from v_i to edge v_j.

3. An adjacency matrix completely defines a simple digraph. . For a loop free graph, the element $a_{ij} = 1, i \neq j$ represents degree of a vertex. The entries along the principal diagonal are all 0's. But, a self loop at the vertex corresponding to $a_{ji} = 1$.

4. Some of the properties of a simple digraph are immediately seen from its adjacency matrix. If a digraph is reflexive, then the diagonal elements of the adjacency matrix are 1's. If a digraph is symmetric, then adjacency matrix is also symmetric.i.e., $a_{ij} = a_{ji} \ \forall\ j$. If a digraph is asymmetric then $a_{ij} = 1 \Rightarrow a_{ji} = 0$ and $a_{ij} = 0 = a_{ji} = 1 \ \forall\ i$ and j.

5. The diagonal elements of A. A^T show that out degree of the vertices. The diagonal elements of A^T. A shows the in degree of the vertices.

6. If G has self loop at every vertex and has no other edge then A(G) is the identity matrix. I_n of order n.

Adjacency matrix can also be used to represent undirected graphs with loops and multiple edges (Pseudo graph). A loop at v_i is represented by a 1 at the $(i, i)^{th}$ position of the adjacency matrix. When multiple edges are present, the adjacency matrix is no longer a zero-one matrix, because the $(i, j)^{th}$ entry of this matrix equals the number of edges that are associated to $\{v_{i,} v_j\}$.

Adjacency matrices for directed multigraphs, on the other hand, are not zero-one matrices, when there are multiple edges in the same direction connecting two vertices. In the adjacency matrix for a directed Multigraph, a_{ij} equals the number of edges that are associated to $(v_{i,} v_j)$.

THEOREM 2.1

Let G be a graph with vertex set $V(G) = \{v_1, v_2, \ldots\ldots, v_n\}$ and adjacency matrix $A = [a_{ij}]$. Then the entry a_{ij}^{k} in row i and column j of A^k is the number of distinct $v_i - v_j$ walks of length k in G.

Proof: We proceed by induction on the no n_i of vertices. Starting with $A = A^1$ we have. each entry $a_{ij} = a_{ji}^{(1)}$, of A either 1 or 0, according to whether $v_i\, v_j$ is or not an edge of G. Hence $a_{ij}^{(1)}$ gives the number of $v_i - v_j$ walks of length 1 in G. Assume, for a positive integer k, that the number $v_i - v_j$ walks of length k in G is given by $a_{ij}^{(k)}$. From matrix multiplication, we have the (i – j) entry $a_{ij}^{(k+1)}$ in A^{k+1} is the dot product of row i of A^k and column j of A. That is,

$$a_{ij}^{(k+1)} =_{ij} a_{it}^{(k)} = a_{i1}^{(k)}a_{1j} + a_{i2}^{(k)}a_{2j} + \ldots\ldots\ldots\ldots + a_{in}^{(k)}a_{nj}\,. \qquad (1)$$

Every $v_i - v_j$ walk W of length k + 1 is produced by beginning with a $v_i - v_t$ walk W^1 of length k for some vertex v_t adjacent to v_j. By induction hypothesis, the number of $v_i - v_t$ walks of length k is a_{it}^{k}, while v_t is adjacent to v_j if and only if $a_{tj} = 1$. Hence by (1), $a_{ij}^{(k+1)}$ does indeed give the number of $v_i - v_j$ walks of length k+1 in G. In view of this theorem, $a_{ii}^{(2)} = \deg v_i$ and $a_{ii}^{(3)}$ is twice the number of triangles in G that contain v_i. Indeed, knowing what entry of A^k represents, allows us to compute powers of adjacency matrices, at least small powers of adjacency matrices of graphs of small orders without actually multiplying matrices.

Example 1: For the graph of Figure 2.4, the adjacency matrix A(G) is

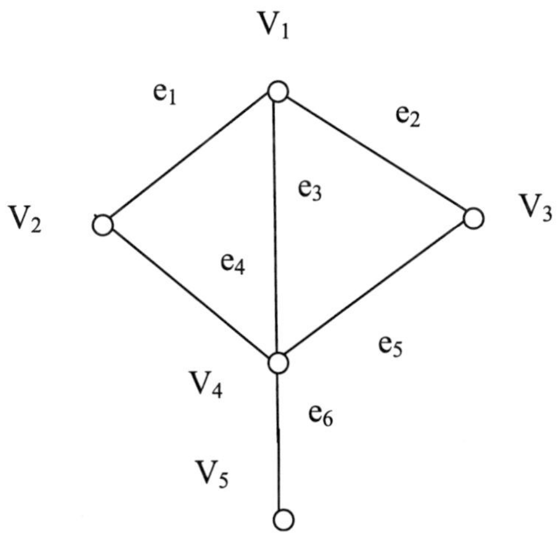

$$A(G) = \begin{pmatrix} 0 & 1 & 1 & 1 & 0 \\ 1 & 0 & 0 & 1 & 0 \\ 1 & 0 & 0 & 1 & 0 \\ 1 & 1 & 1 & 0 & 1 \\ 0 & 0 & 0 & 1 & 0 \end{pmatrix}$$

Figure 2.4 Adjacency matrix

$$A^2 = \begin{pmatrix} 3 & 1 & 1 & 2 & 1 \\ 1 & 2 & 2 & 1 & 1 \\ 1 & 2 & 2 & 1 & 1 \\ 2 & 1 & 1 & 4 & 0 \\ 1 & 1 & 1 & 0 & 1 \end{pmatrix}$$

$$A^3 = \begin{pmatrix} 4 & 5 & 5 & 6 & 2 \\ 5 & 2 & 2 & 6 & 1 \\ 5 & 2 & 2 & 6 & 1 \\ 6 & 6 & 6 & 4 & 4 \\ 2 & 1 & 1 & 4 & 0 \end{pmatrix}$$

Powers of adjacency matrix

For the graph G of Figure 2.4, the entry $a_{41}^{(3)}$of A^3 is the number of distinct $v_4 - v_1$ walks in G of length 3. Since there are six $v_4 - v_1$ walks of length 3 in G, namely:

(1) $v_{4,} v_{5,} v_4, v_1$

(2) $v_{4,} v_{2,} v_4, v_1$

(3) $v_4, v_{3,} v_4, v_1$

(4) $v_{4,} v_{1,} v_2, v_1$

(5) $v_4, v_1, v_{3,} v_1$

(6) v_4, v_1, v_4, v_1. It follows that $a_{41}^{(3)} = 6$, as we have seen.

The next result immediately follows from the above theorem.

Corollary 2.2

Let G be a graph with n vertices $v_1, v_2, \ldots\ldots, v_n$ and let A denote the adjacency matrix of G with respect to this listing of the vertices. Let $B = b_{Ij}$ be the matrix.

$$B = A + A^2 + A^3 + \ldots\ldots\ldots + A^{n-1}.$$

Then G is connected graph iff for every pair of distinct indices i and j, we have $b_{Ij} \neq 0$. i.e., if B has no zero entries off the main diagonal. We can use this result to check whether a graph is connected or not.

Example: 2

A graph G has the following adjacency matrix. Check whether it is connected?

$$A(G) = \begin{pmatrix} 0 & 0 & 1 & 0 & 0 \\ 0 & 0 & 0 & 1 & 0 \\ 0 & 0 & 0 & 0 & 1 \\ 0 & 1 & 0 & 0 & 1 \\ 0 & 0 & 1 & 1 & 0 \end{pmatrix}$$

Solution

Since here n = 5, we look at $B = A + A^2 + A^3 + A^4$. We find A^2, A^3, A^4 by matrix multiplication.

$$A^2 = \begin{pmatrix} 1 & 0 & 0 & 0 & 1 \\ 0 & 1 & 0 & 0 & 1 \\ 0 & 0 & 2 & 1 & 0 \\ 0 & 0 & 1 & 2 & 1 \\ 1 & 1 & 0 & 0 & 2 \end{pmatrix} \quad A^3 = \begin{pmatrix} 0 & 0 & 2 & 1 & 0 \\ 0 & 0 & 1 & 2 & 0 \\ 2 & 1 & 0 & 0 & 3 \\ 1 & 2 & 0 & 0 & 3 \\ 0 & 0 & 3 & 3 & 0 \end{pmatrix}$$

$$\text{Now } B = A^2 + A^3 + A^4 = \begin{pmatrix} 3 & 1 & 3 & 1 & 4 \\ 1 & 3 & 1 & 3 & 4 \\ 3 & 1 & 7 & 5 & 4 \\ 1 & 3 & 5 & 7 & 4 \\ 4 & 4 & 4 & 4 & 8 \end{pmatrix}$$

Since B has no non zero entry of the main diagonal, the graph is connected.

Another common way to represent graph is to use Incidence matrix.

2.2 INCIDENCE MATRIX

As an alternative to the adjacency matrix, it is sometimes more convenient to represent a graph by its incidence matrix. Suppose that G has n vertices listed as v_1, v_2, ... v_n and t edges listed as e_1, e_2, ...e_t. The incidence matrix of G is the n × m matrix $M(G) = (m_{ij})$ where m_{ij} is the number of times that the vertex v_i is incident with the edge e_j. i.e., $m_{ij} = 1$ if vertex v_i is incident on edge e_j.

$$m_{ij} = \begin{cases} 1 & \text{if vertex } v_i \text{ is incident on edge } e_j \\ 0 & \text{otherwise} \end{cases}$$

The matrix has a row for every vertex and a column for every edge. The non zero elements in the matrix is equal to the sum of degrees of all the vertices of the graph. Sum of the elements in the ith row of M(G) gives is the degree of the vertex v_i. Figure 2.5 specifies 5 × 8 matrix representation for the graph.

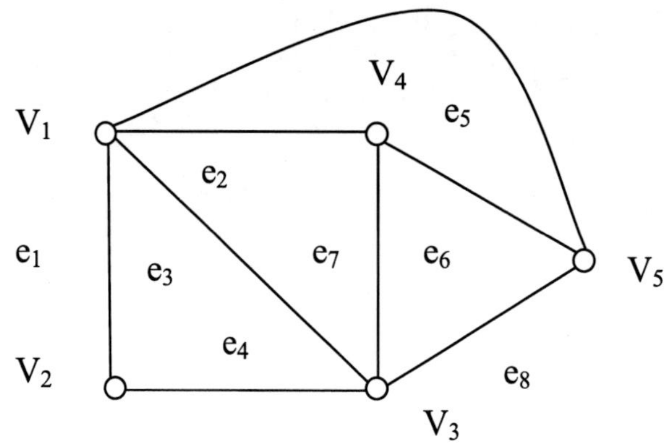

Figure 2.5

$$M = \begin{array}{c} \\ v_1 \\ v_2 \\ v_3 \\ v_4 \\ v_5 \end{array} \begin{array}{c} \begin{array}{cccccccc} e_1 & e_2 & e_3 & e_4 & e_5 & e_6 & e_7 & e_8 \end{array} \\ \begin{pmatrix} 1 & 1 & 1 & 0 & 1 & 0 & 0 & 0 \\ 1 & 0 & 0 & 1 & 0 & 0 & 0 & 0 \\ 0 & 0 & 1 & 1 & 0 & 0 & 1 & 1 \\ 0 & 0 & 0 & 0 & 1 & 1 & 0 & 1 \\ 0 & 1 & 0 & 0 & 0 & 1 & 1 & 0 \end{pmatrix} \end{array}$$

Incidence Matrix

We make the following observations about the incidence matrix M of a graph G.

1. Each column contains exactly two unit elements (i.e., it contains only 0 and 1).
2. A row with all zero elements corresponds to an isolated vertex.
3. A row with a single unit element corresponds to a pendent vertex.
4. The number of 1's in each row equals to the degree of the corresponding vertex.
5. Every edge is incident on exactly two vertices, each column of the incidence matrix has exactly two ones except the loop. Loops are represented using a column with exactly one entry equal to 1, corresponding to the vertex that is incident with this loop.
6. The parallel edges in a graph produce identical columns in its incidence matrix, because these edges are incident with the same pair of vertices.
7. Permutation of any two rows or columns in an incidence matrix simply corresponds to relabeling the vertices and the edges of the same graph.

For a Digraph, G consists of n vertices and m edges. An n × m incidence matrix, $M = [m_{ij}]$ is defined as

$$M_{ij} = \begin{cases} 1 & \text{if } v_i \text{ is initial vertex of edge } e_j \\ -1 & \text{if } v_j \text{ is final vertex of edge } e_j \\ 0 & \text{if } v_i \text{ is not incident on edge } e_j \end{cases}$$

The non zero elements in the matrix is equal to the number of edges in the Digraph. The following is the 5 × 7 incidence matrix in Figure 2.6 is

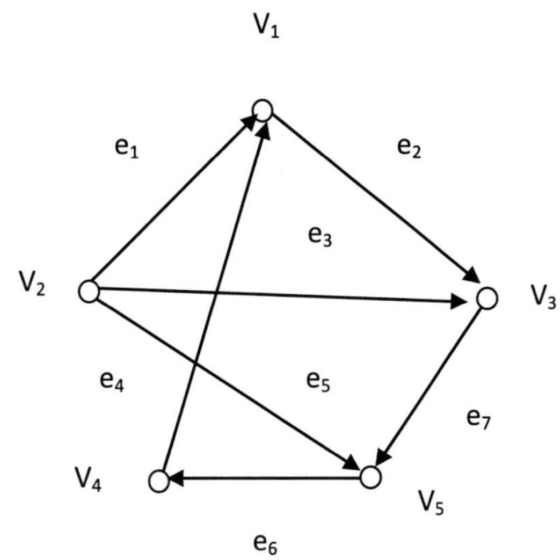

Figure 2.6

M =

	e_1	e_2	e_3	e_4	e_5	e_6	e_7
v_1	-1	1	0	-1	0	0	0
v_2	1	0	1	0	1	0	0
v_3	0	-1	-1	0	0	0	1
v_4	0	0	0	1	0	-1	0
v_5	0	0	0	0	-1	1	-1

Incidence Matrix

Another example is the well known Ecological system, where in the node list has in addition to plants and animals, as other factors like soil type, weather condition, temperature, fertilizers, disease and pesticides and after generation of the node list, form relationship between related nodes. The links among them are represented by an n incidence matrix, where n is the number of nodes. Firstly, the value of an entry representing the link between nodes i and j is set at 1, if there is an interaction between the species / environmental factors represented by these nodes. Otherwise, the value of the matrix entries is set to zero. In addition, one adds + or – sign to the links represented by 1, expressing the nature of the link. A link from A to B is given +1 if the presence of species / environmental factor A facilitates the presenceof species/environmental factor B. The value is set –1 for the reverse case. If A consumes B, then there exists a link from A to B weighted with +1, and there alsoexists a link from B to A weighted with -1, to denote that B is eaten by A. For example, for the simple network of ecosystem represented in Figure 1. 8, the +1 weighted link is denoted by the direction of the arrow, and the –1 weighted link by the opposite direction. (eg., a bird feeds on slug and the slug is consumed by the bird). To describe the changes in the development of the ecological system, four interaction matrices corresponding to four seasons are in this way prepared. Similarly interaction matrices as regards soil classification, pesticides, fertilizers, diseases and weather conditions etc., are prepared to assess the impact of different stressors (by removing the concerned damaged node from the network) affecting the ecological system.

2.3 FUSION OF GRAPHS

Let u and v be distinct vertices of a graph G. We can construct a new graph G_1 by fusing (or identifying) the two vertices by replacing them with a single new vertex x such that every edge that was incident with either u or v in G is now incident with x i.e., the end u and the end v become end x. Thus the new graph has one less vertex than G but the same number of edges as G and the degree of the vertex x is sum of degrees of u and v.

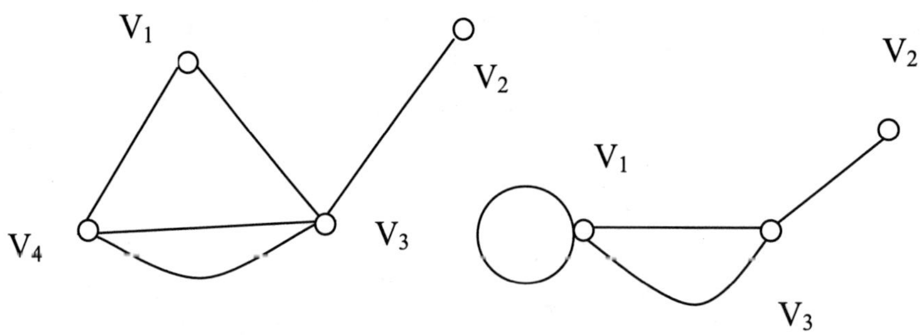

Figure 2.7 v_1 fused with v_4

To find the adjacency matrix after two adjacent vertices u and v are fused, we require the following two steps:

Algorithm 1 To find the new adjacency matrix after fusion

Step 1: Change u's row to the sum of u's row and change (symmetrically) change u's column to the sum of u's column with v's column.

Step2: Delete the row and column corresponding to v. The resulting matrix is the adjacency matrix of the new graph G.

We hereby give an alternative way of finding whether a graph is connected or not by using fusion. This is a better method as the earlier method involves too much of matrix multiplication.

Algorithm 2 Fusion algorithm for connectedness.

Step 1: Replace G by its underlying simple graph. To get the adjacency matrix of new graph just replace all non zero entries off the main diagonal by 1 and make all entries on the diagonal 0. Denote the underlying simple graph also as G.

Step 2: Fuse vertex v_1 to the first of the vertices $v_2, \ldots\ldots\ldots, v_n$ with which it is adjacent to give a new graph, also denoted by G, in which the new vertex is also denoted by v_i.

Step 3: The above two step process gives the adjacency matrix A(G).

Step 4: Repeat steps 1 and 2 with v_1 until v_1 is not adjacent to any of the other vertices'.

Step 5: Repeat step 2 and 4 on the vertex v_2 of the last graph and then on all remaining vertices of the resulting graphs.

The final graph is empty and the number of its (isolated) vertices is the number of connected components of the initial graph G. We give an example to illustrate the above algorithm.

Example: 3

Given below is the adjacency matrix of graph G with seven vertices listed as $v_1, v_2, \ldots\ldots\ldots, v_7$. Use fusion algorithm to check the connectedness.

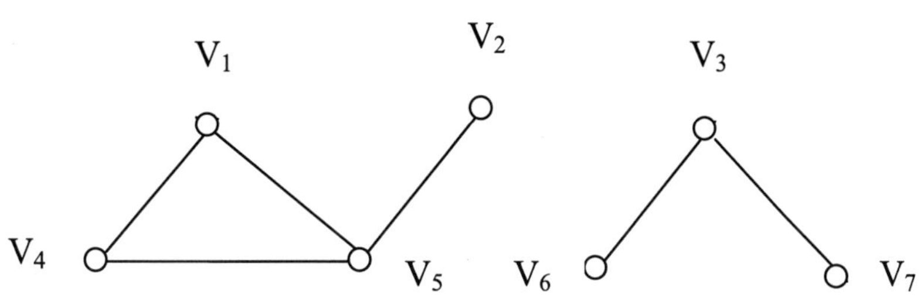

Figure 2.8 Graph G

Solution

$$A(G) = \begin{pmatrix} 0 & 0 & 0 & 1 & 1 & 0 & 0 \\ 0 & 0 & 0 & 0 & 1 & 0 & 0 \\ 0 & 0 & 0 & 0 & 0 & 2 & 1 \\ 1 & 0 & 0 & 2 & 0 & 0 & 0 \\ 1 & 1 & 0 & 2 & 0 & 0 & 0 \\ 0 & 0 & 2 & 0 & 0 & 0 & 0 \\ 0 & 0 & 1 & 0 & 0 & 0 & 0 \end{pmatrix}$$

Adjacency matrix of graph G

Step 1: Replacing G by its underlying simple graph

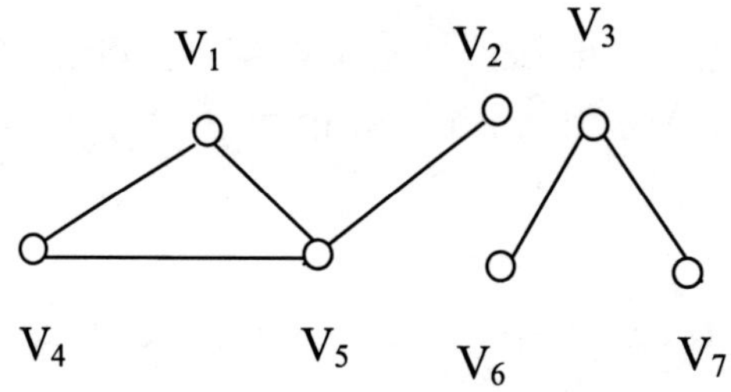

$$A(G) = \begin{pmatrix} 0 & 0 & 0 & 1 & 1 & 0 & 0 \\ 0 & 0 & 0 & 0 & 1 & 0 & 0 \\ 0 & 0 & 0 & 0 & 0 & 1 & 1 \\ 1 & 0 & 0 & 0 & 1 & 0 & 0 \\ 1 & 1 & 0 & 1 & 0 & 0 & 0 \\ 0 & 0 & 1 & 0 & 0 & 0 & 0 \\ 0 & 0 & 1 & 0 & 0 & 0 & 0 \end{pmatrix}$$

Fusing v_1 with v_4

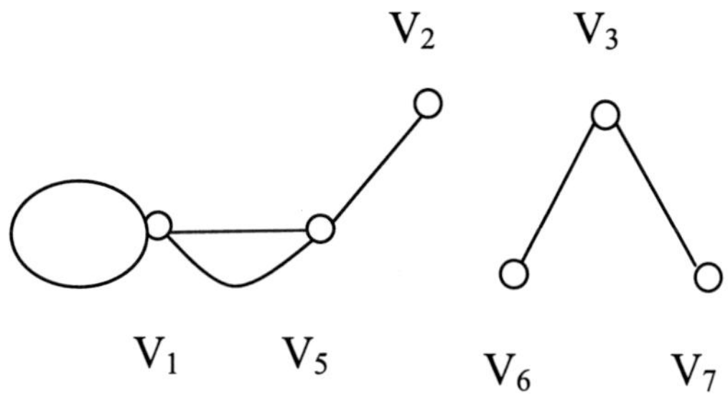

$$A(G) = \begin{pmatrix} 1 & 0 & 0 & 2 & 0 & 0 \\ 0 & 0 & 0 & 1 & 0 & 0 \\ 0 & 0 & 0 & 0 & 1 & 1 \\ 2 & 1 & 0 & 0 & 0 & 0 \\ 0 & 0 & 1 & 0 & 0 & 0 \\ 0 & 0 & 1 & 0 & 0 & 0 \end{pmatrix}$$

Repeat step 1 on graph obtained in step 2.

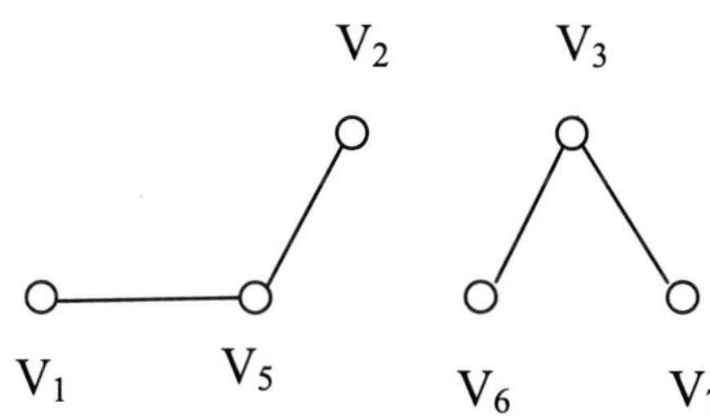

$$\begin{pmatrix} 0 & 0 & 0 & 1 & 0 & 0 \\ 0 & 0 & 0 & 1 & 0 & 0 \\ 0 & 0 & 0 & 0 & 1 & 1 \\ 1 & 1 & 0 & 0 & 0 & 0 \\ 0 & 0 & 1 & 0 & 0 & 0 \\ 0 & 0 & 1 & 0 & 0 & 0 \end{pmatrix}$$

Fusing v_1 with v_5

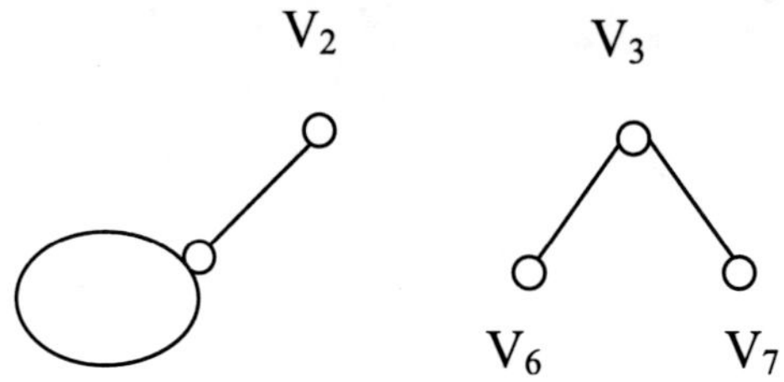

$$\begin{pmatrix} 1 & 1 & 0 & 0 & 0 \\ 1 & 0 & 0 & 0 & 0 \\ 0 & 0 & 0 & 1 & 1 \\ 0 & 0 & 1 & 0 & 0 \\ 0 & 0 & 1 & 0 & 0 \end{pmatrix}$$

Repeating step 1

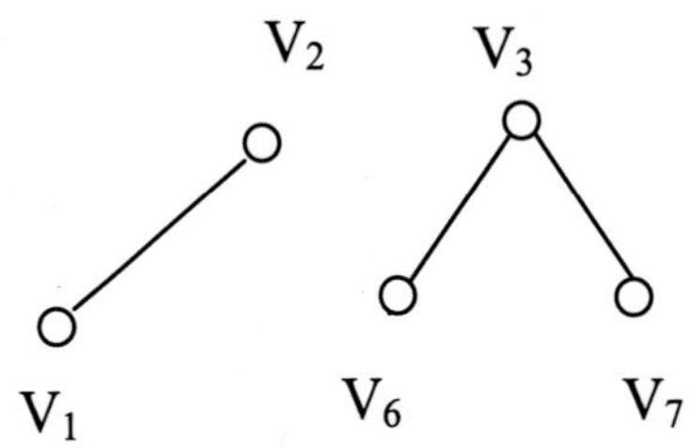

$$\begin{pmatrix} 0 & 1 & 0 & 0 & 0 \\ 1 & 0 & 0 & 0 & 0 \\ 0 & 0 & 0 & 1 & 1 \\ 0 & 0 & 1 & 0 & 0 \\ 0 & 0 & 1 & 0 & 0 \end{pmatrix}$$

Fusing v_1 with v_2,

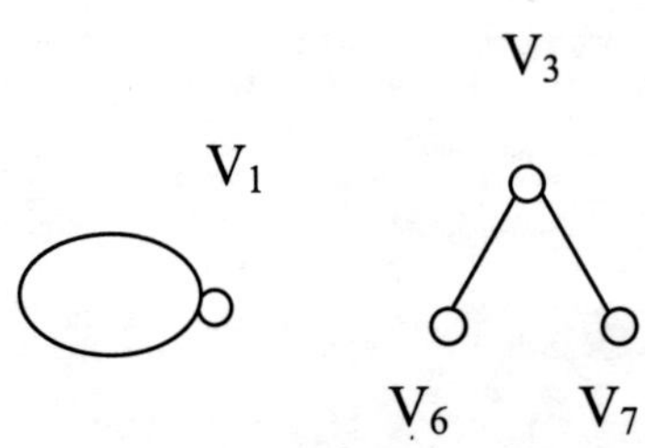

$$\begin{pmatrix} 1 & 0 & 0 & 0 \\ 0 & 0 & 1 & 1 \\ 0 & 1 & 0 & 0 \\ 0 & 1 & 0 & 0 \end{pmatrix}$$

Repeating step 1,

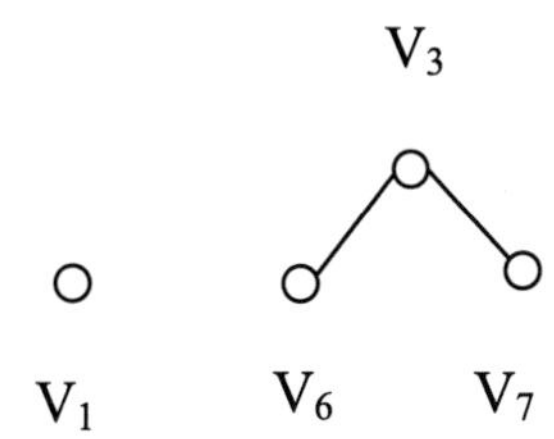

$$\begin{pmatrix} 0 & 0 & 0 & 0 \\ 0 & 0 & 1 & 1 \\ 0 & 1 & 0 & 0 \\ 0 & 1 & 0 & 0 \end{pmatrix}$$

Fusing v_3 with v_6,

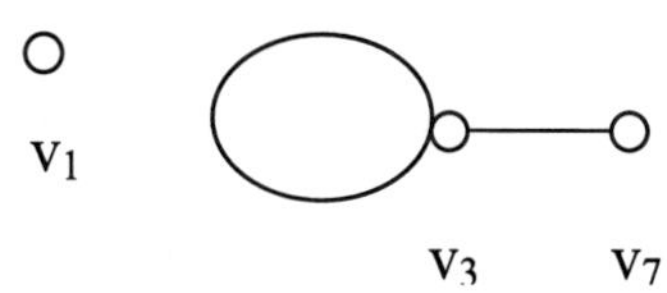

$$\begin{pmatrix} 0 & 0 & 0 \\ 0 & 1 & 1 \\ 0 & 1 & 0 \end{pmatrix}$$

Repeating step 1,

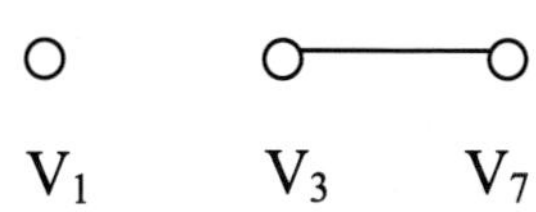

$$\begin{pmatrix} 0 & 0 & 0 \\ 0 & 0 & 1 \\ 0 & 1 & 0 \end{pmatrix}$$

Fusing v_3 with v_7,

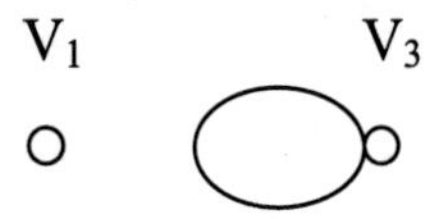

$$\begin{pmatrix} 0 & 0 \\ 0 & 1 \end{pmatrix}$$

Repeating step 1

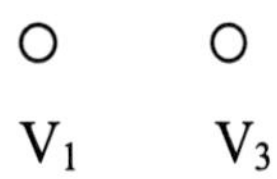

$$\begin{pmatrix} 0 & 0 \\ 0 & 0 \end{pmatrix}$$

Since the final adjacency matrix is a 2×2 null matrix, we conclude that original graph G has 2 connected components.

2.4 GRAPH ISOMORPHISM

We often need to know when two graphs are essentially the same and when they are essentially different. (that is, whether it is possible to draw two graphs in the same way). For instance, in chemistry, graphs are used to model compounds. Different compounds can have the same molecular formula but can differ in structure. Such compounds will be represented by graphs that cannot be drawn in the same way. The graphs representing previously known compounds can be used to determine whether supposedly new compound has been studied before

When we say graphs are essentially the same, we mean that they differ in the way they are drawn or labeled. There should be a one-to-one correspondence between the vertices of the graphs (which preserves adjacency) and one-to-one correspondence between their edges such that corresponding vertices are incident with corresponding edges. Indeed, there is a terminology for graphs with the same structure. i. e. Graphs with same vertices and edges but differ only in the way of representation. The proper term for "essentially the same" is Isomorphic.

Two graphs G_1 and G_2 are isomorphic (written $G_1 \cong G_2$ or sometimes $G_1 \equiv G_2$) if there is a function φ: $V(G_1) \rightarrow V(G_2)$ from the vertices of G_1 to the vertices of G_2 such that (i) φ is one-to-one (ii) φ is onto and (iii) For each pair of vertices u, and v of G $\{u, v\} \in E(G) \leftrightarrow \{\varphi(u), \varphi(v) \in E(G_2)\}$. We call φ an isomorphism from G_1 to G_2 and, abusing notation, say that $\varphi : G_1 \rightarrow G_2$ is an isomorphism. It is basically a possible redrawing of the graphs. For example, G_1 and G_2 of Figure 2. 9 are isomorphic under the correspondence $\varphi(u_1) = u_2$, $\varphi(v_1) = w_2$, $\varphi(w_1) = y_2$, $\varphi(x_1) = v_2$, $\varphi(y_1) = x_2$ and so $G_1 \cong G_2$. That is, $G_1 \cong G_2$, if the vertices of G_1 and G_2 can be labeled (or relabeled) without changing any of the incidence relations to produce two equal graphs.

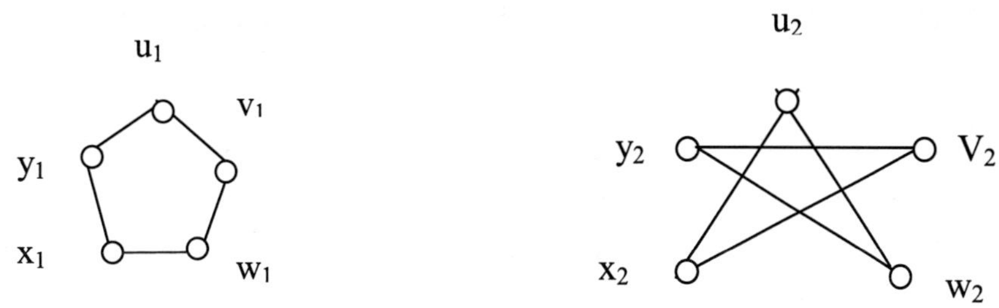

(a) (b)

Figure 2.9

The above definition for isomorphism is valid for digraph also. As expected, two digraphs D and D'are isomorphic, written as $D \cong D'$, if there exists a **one-to-onecorrespondence**: $\varphi = V(D) \rightarrow V(D')$, such that $(u, v) \in E(D)$ if and only if $(\varphi(u), \varphi(v) \in (E(D')$. The function φ is called isomorphism.

Suppose: $G_1 \rightarrow G_2$is an isomorphism from a graph G_1 to another graph G_2. If v is a vertex of degree k in G_1 and if $v_1, v_2, \ldots\ldots\ldots, v_k$ are the vertices adjacent to v, then, in $G_{2,}$) is adjacent to k vertices $\varphi(v_1), \varphi(v_2), \ldots\ldots, \varphi(v_k)$, but to no other vertex. Thus, the degree of $\varphi(v)$ is also k. It follows that isomorphic graphs have the same degree sequences and hence also the same number of edges, since the number of edges in a graph is one half the sum of the vertex degrees. That is, not only are the orders the same and the sizes the same of two isomorphic graphs, so too are the degrees of their vertices. This observation provides us with the following theorem.

THEOREM 2.3

If G_1 and G_2 are isomorphic graphs, then the degrees ofthevertices of G_1 are the same as the degree of the vertices of G_2.

Proof: Since G_1 and G_2 are isomorphic, there is an isomorphism: $\varphi : V(G_1) \rightarrow V(G_2)$. Let u be a vertex of G_1 and suppose that $\varphi(u) = v$, where v, therefore belongs to G_2. We show that the degree of v in G_2 is the same as degree of u in G_1. Suppose first that the $\deg_{G1} u = 0$. Then no vertex of G_1 is adjacent to u. Let y be any vertex in G_2 that is different from v. Then there is a vertex x in G_1 such that $\varphi(x) = y$, since x and u are not adjacent in G, the vertices y and v are not adjacent in G_2. So no vertex of G_2 is adjacent to v and $\deg G_2 v = 0$. Suppose next that $\deg_{G1} u = k \geq 1$. Let $N(u) = \{x_1, x_2, \ldots\ldots x_k\}$ be the set of vertices adjacent to u in G, where $\varphi(x_i) = y_i$ for $1 \leq i \leq k$, Since u and x_i are adjacent for $1 \leq i \leq k$. the vertices v and y_i are adjacent for $1 \leq i \leq k$. If y is a vertex of G_2 such that $\varphi \in \{v\} \cup \{y_1, y_2, \ldots\ldots, y_k\}$, then there is a vertex x in G_1 such that $\varphi(x) = y$, where $x \in \{u\} \cup N(u)$. Since x and u are not adjacent in G, the vertices v and y are not adjacent in G_2. So $\deg G_2\, v = k$.

The theorem 2.3, not only tells us that two isomorphic graphs G_1 and G_2 have the same degree sequence but also says that if is an isomorphism from $V(G_1)$ to $V(G_2)$ and u is a vertex of G_1 then

$deg_{G1}\,u = deg_{G2}$ that is, under an isomorphism a vertex can only map into a vertex having the same degree.

In general, isomorphic graphs have the same degree sequences and hence also the same number of edges, since the number of edges is one half the sum of the vertex degrees. But unfortunately, two graphs may have the same degree sequence (and hence the same number of vertices and edges) and still not be isomorphic. In this case, we must dig deeper for reasons why they are not isomorphic. In general terms; one approach is to classify the vertices into classes according to some property preserved by isomorphism. We now make all of this more precise.

Recall that Isomorphism produces a relation on any set of graphs, namely, a graph G_1 is related to a graph G_2 if G_1 is isomorphic to G_2. This relation is an equivalence relation. This says that isomorphism is Reflexive, Symmetric, and Transitive.

We have remarked that the definition of isomorphismis

Symmetric: $G_1 \cong G_2$ if and only if $G_2 \cong G_1$.

It is also **reflexive:** G $\cong$ G for any graph. (every graph is isomorphic to itself, because the identity map $G_1 \rightarrow G_2$ is an isomorphism)

Transitive: if $G_1 \cong G_2$ and $G_2 \cong G_3$, then $G_1 \cong G_3$.

Thus isomorphism is an ***equivalence relation*** onthe set of all graphs. One of the major consequence of knowing that isomorphismis an equivalence relation on a set of graphs is that this produces a partition of this set into equivalence classes. That is, the set of all graphs is therefore partitioned into disjoint equivalence classes (subsets) called isomorphism classes.

Any two graphs in the same equivalence class are isomorphic (Figure 2.10), two graphs in different equivalence classes are not isomorphic (Figure 2.11). When we usually remark that two graphs are different, we really mean "in different isomorphism classes".

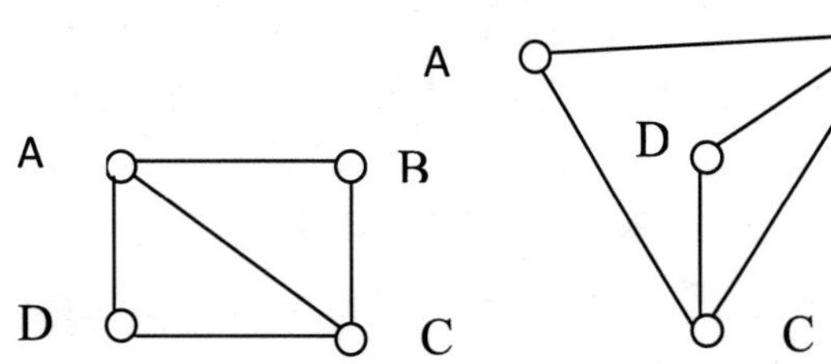

Figure 2.10 Two Isomorphic graphs

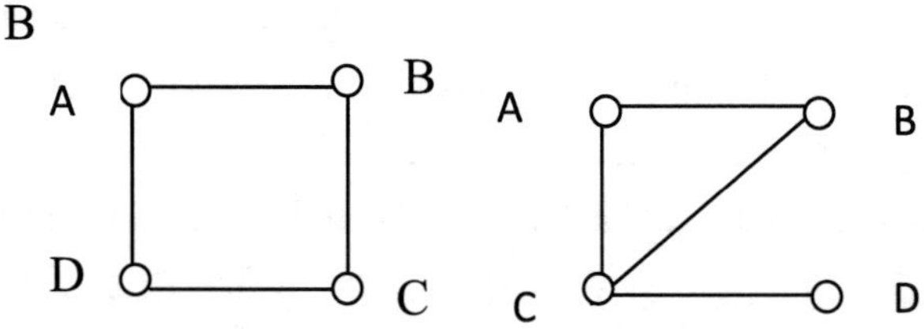

Figure 2.11 Two Non isomorphic Graphs

Another simpler way to verify two graphs $G_1$and $G_2$to be isomorphic, is to involve the use of complements of graphs. That is the observation that "***Two graphs G_1 and G_2 are isomorphic, if and only if, their complements $G_1^c$and G_2^c are isomorphic***". This follows directly from the definition of isomorphism and complement. A graph and its complement may, in fact, be isomorphic to each other.

On the other hand, Suppose that G_1 and G_2 are two graphs and that $\varphi : V(G_1) \rightarrow V(G_2)$ is a one to one on to function. Let A be the adjacency matrix for the vertex ordering $v_1, v_2, \ldots, v_n$ of the vertices of G_1 .Let A' be the adjacency matrix for the vertex ordering $\varphi(v_1), \ldots, \varphi(v_n)$. Then φ, is an isomorphism from $V(G_1)$ to $V(G_2)$ if and only if the adjacency matrices A and A' are equal.

But, if the adjacency matrices A and A' are not equal, then all that proves is that the function itself is not an isomorphism. it may still be the case that graphs G_1 and G_2 are isomorphic under some other function. The fact that an isomorphism must map a vertex of G_1 of degree d to a vertex of G_2 of degree d, will shorten the search for isomorphism. . .

Often it is quite difficult to determine whether two graphs are isomorphic or not. There are n possible one-to-one correspondence between vertex sets of two simple graphs with n vertices. Testing each such correspondence to see whether it preserves adjacency and non adjacency is impractical if n is large. The problem becomes even harder as the number of vertices and edges of the graph gets larger.

We can show that two graphs are not isomorphic by showing that they do not share a property that isomorphic graphs must have. Such property is called an invariant with respect to the isomorphism of graphs. That is, a property preserved by isomorphism of graphs is called a graph invariant. A complete set of invariants determines a graph up to isomorphism The invariants are

(i) The number of vertices

(ii) The number of edges

(iii) And the degree sequence of the two graphs.

If any of these quantities differ in two graphs, those graphs can't be isomorphic. However, when these invariants are the same it doesn't mean that two graphs are isomorphic. Apart from these invariants, we need a one-to-one and onto function which preserves adjacency of the vertices in simple graph and preserves the direction of edges in digraphs. There are no useful sets of invariants currently known that can be used to determine whether simple graphs are isomorphic. Finding a simple and efficient criterion of isomorphism is still actively pursued and is an important unsolved problem in graph theory.

SUMMARY

1. The adjacency matrix A (G) of a finite graph G on n vertices is the n × n matrix in hich the entry aij is the number of edges in G with end points (v_1, v_j). The adjacency matrix foran undirected graph is symmetric and that for an Isomorphism classof graphs is unique.

2. The incidence matrix M(G) ofa finite graph G is the $n \times m$ *matrix* in which entry m_{ij} is 1 if the vertex v_i is an end point of e_i and otherwise 0.

3. Two Graphs G And H are isomorphic if the vertices of one can be rearranged without adding or deleting any edges, so that the two graphs are identical ignoring the labels on the vertices.

EXCERCISES

1. Represent the given graph using adjacency matrix

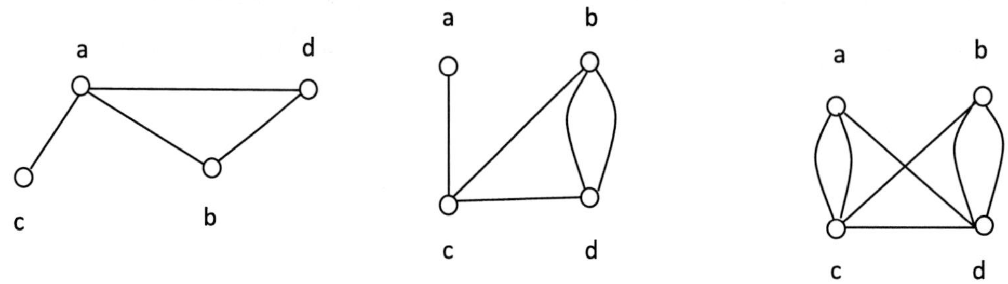

2. Draw the graphs with the given adjacency matrix

$$(a)\begin{pmatrix} 0 & 0 & 1 & 1 \\ 0 & 0 & 1 & 0 \\ 1 & 1 & 0 & 1 \\ 1 & 1 & 1 & 0 \end{pmatrix} \qquad (b)\begin{pmatrix} 1 & 2 & 0 & 1 \\ 2 & 0 & 3 & 0 \\ 0 & 3 & 1 & 1 \\ 1 & 0 & 1 & 0 \end{pmatrix}$$

$$(c)\begin{pmatrix} 0 & 0 & 1 & 1 \\ 0 & 0 & 1 & 0 \\ 1 & 1 & 0 & 1 \\ 1 & 1 & 1 & 0 \end{pmatrix} \qquad (d)\begin{pmatrix} 1 & 2 & 0 & 1 \\ 2 & 0 & 3 & 0 \\ 0 & 3 & 1 & 1 \\ 1 & 0 & 1 & 0 \end{pmatrix}$$

3. Use incidence matrix to represent the graph?

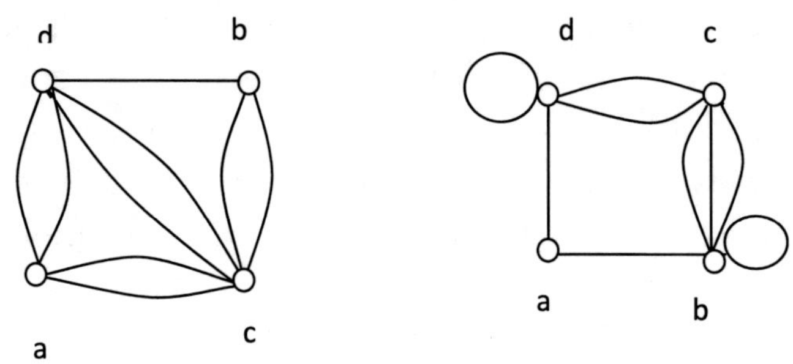

4. For each of the ten pairs of graphs that can be obtained from those shown, either label the graphs soas to exhibit an isomorphism or explain why the graphs are not isomorphic? Draw all non isomorphic graphs on 3 vertices. Give the degree sequence of each?

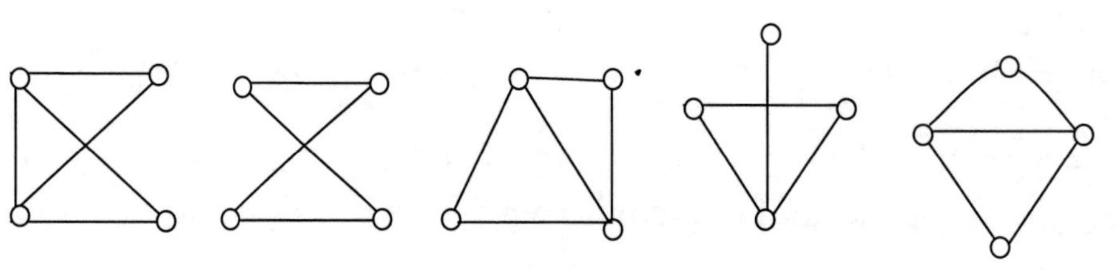

(i) ii) (iii) (iv) (v)

5. Give an example of an three non isomorphic graph of order 5 and sixe 5?

6. For each of the pair of graphs listed below identify whether they are Isomorphic

(i) 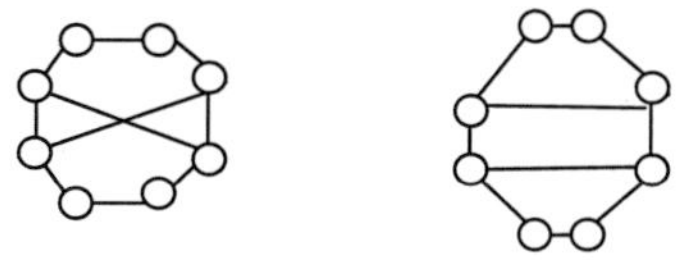

(ii) 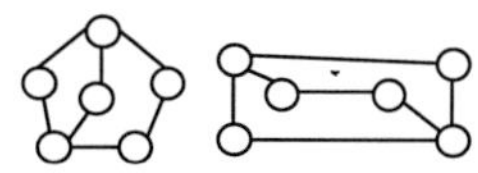

(iii) 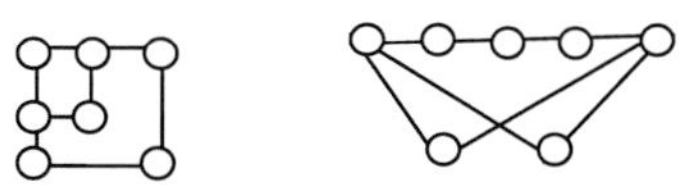

(iv) 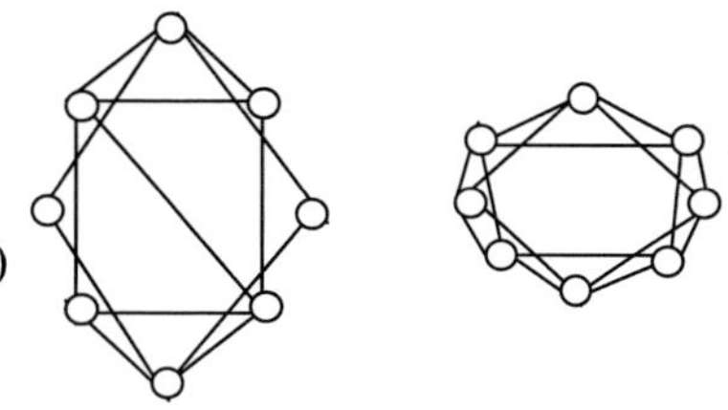

(v) 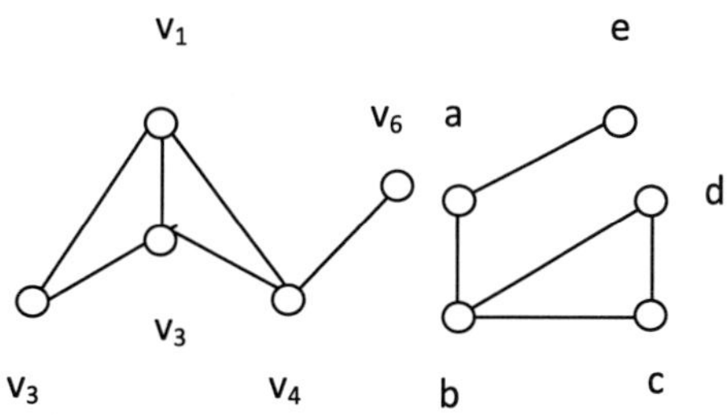

7. Use fusion algorithm to determine whether the graph specified by the following adjacencymatrixis connected or not?

$$\begin{pmatrix} 0 & 1 & 2 & 3 \\ 1 & 0 & 3 & 2 \\ 2 & 3 & 0 & 1 \\ 3 & 2 & 1 & 0 \end{pmatrix} \text{ and } \begin{pmatrix} 0 & 0 & 1 & 0 \\ 0 & 0 & 0 & 1 \\ 0 & 1 & 0 & 0 \\ 0 & 0 & 1 & 0 \end{pmatrix}$$

8. Is the graph with the followingadjacency matrices connected

$$A(G) = \begin{pmatrix} 0 & 0 & 1 & 0 & 0 \\ 0 & 0 & 0 & 1 & 0 \\ 0 & 0 & 0 & 0 & 1 \\ 0 & 1 & 0 & 0 & 1 \\ 0 & 0 & 1 & 1 & 0 \end{pmatrix} \text{ and } A(G) = \begin{pmatrix} 1 & 1 & 0 & 1 & 1 \\ 1 & 1 & 1 & 0 & 1 \\ 1 & 1 & 1 & 1 & 0 \\ 1 & 0 & 1 & 1 & 0 \\ 1 & 1 & 0 & 0 & \end{pmatrix}$$

3

PATHS AND CIRCUITS

3.0 INTRODUCTION

Many of the concepts that occur in graph theory and which we investigate in detail concern various ways in which one can "move about" in a graph. In Particular, if we think of the vertices of the graph as locations and the edges as links between certain pairs of locations, then the graph can be considered to be as modeling some community. There is a variety of kinds of trips that can be taken in the community. The concepts such as paths, circuits, and Euler graphs deal mainly with the connectivity in graphs. Obviously, one important application of graph theory is in the area of puzzles and games. Often a puzzle can be converted into a graphical problem to determine the existence or non existence of an ' Eulerian trail" or a Hamiltonian cycle within the graph, is discussed in the first section of this chapter. As mentioned in Chapter 1, the concept of Euler graphs was formulated when Euler studied the problem of Konigsberg Bridge problem. The study of the Hamiltonian Graphs, on the other hand was initiated in the 19th century. Both these graph possess rich structure and their study is a very active field of research for graph theorists. Several structural theorems of these graphs are also discussed in the later sections. Also we discuss two algorithms to find the Euler cycle and Hamiltonian path (of the complete weighted graph).

3.1 WALKS, PATHS AND CYCLES

A walk in a graph G is a finite sequence

$\mathbf{W = v_0 e_1, v_1 e_2, \ldots\ldots\ldots\ldots, v_{k-1} e_k v_k.}$

Whose terms are alternatively vertices and edges such that, for $1 \leq i \leq k$, the edge e_i has ends v_{i-1} and v_i. The vertex v_o is called the origin and v_k the ***terminus*** of walk W. They need not be distinct. Vertices $v_1 \ldots\ldots v_{k-1}$ are called ***internal vertices***. The number of edges in W is called ***length of the walk***. Vertices and edges can be repeated in a walk. The process of joining two walks which have a common end point to form a longer walk is called ***concatenation*** (stringing together). A ***trivial walk*** is one containing no edges and has length zero . A walk is ***closed; open*** depending on whether origin and terminus are same or different. If the walk is closed, then it is a ***cycle*** provided its n points are distinct and $n \geq 3$. If the edges $e_1, e_2, \ldots\ldots\ldots\ldots e_k$ of Walk W are distinct then we call such a walk as ***trail***. So a ***trail is a walk if no edge is repeated***. If the vertices $v_0, v_1, \ldots\ldots, v_k$ of the walk $W = v_0e_1\, v_1e_2, \ldots\ldots\ldots\ldots, v_{k-1}\, e_k\, v_k$. are distinct then W is called a path. The path is any route taken along the edges of the graph. A ***simple path*** is the one in which no edge is repeated in the walk. That is, it is also a trail where all the vertices and all edges are different except the first vertex is equal to the last vertex. An edge which is not a loop is a path of length one. A loop can be included in a walk but not in a path. In a path the terminal vertices are of degree 1 when other vertices (intermediate vertices) are of degree 2. A path is a walk if no vertex is repeated. ***So every path is a trail(open walk) and not every trail is a path.*** When writing a path, we usually omit the edges. .

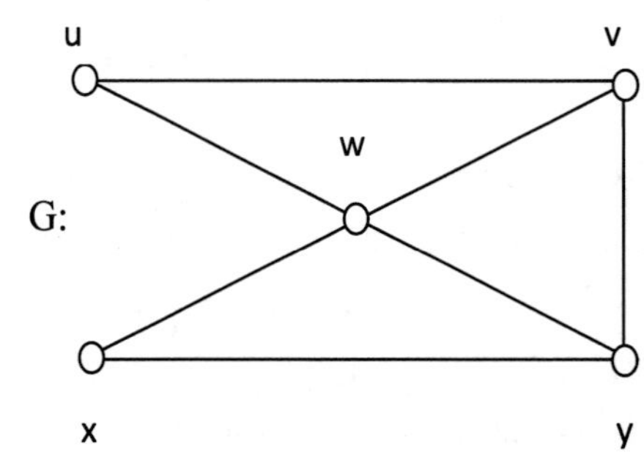

Figure 3.1 Showing Walks in a Graph

For the graph of Figure 3. 1, W : x, y, w, y, v, w is a walk of length 5. A walk of length 0 is a trivial walk. So W: v is a trivial walk. The sub graphs G_1, G_2, G_3, G_4 of the graph G are a trail, path, circuit and cycle respectively. The x-w walk in W : x, y, w, y, v, w is not an x-w trail as edge wy is repeated. On the other hand, T:u, w, y, x, w, v is a u-v trail in the graph of G. Notice that the trail repeats vertex w. Although the definition of a trail stipulates that no edge can be repeated, no such condition is placed on the vertices. A u-v walk in a graph in which no vertices are repeated is a u-v path. So P : u, w, y, v is a u-v path, while the u-v trail T:u, w, y, x, w, v is not a u-v path.

A nontrivial closed trail in a graph is called a ***cycle***, If its origin and internal vertices are distinct. The closed trail $C = v_1v_2 \ldots\ldots\ldots\ldots v_n$ is a cycle if C has at least one edge and $v_1v_2 \ldots\ldots\ldots\ldots v_n$ are n distinct vertices. The term cycle will always imply the presence of at least 3 distinct edges. A cycle with K edges is called a ***K-cycle***. A cycle, unlike a path, is not allowed to have length 0. A K-cycle is called odd or even depending on whether K is odd or even. Any two cycles with same length are isomorphic. An n cycle is denoted by C_n We can define some cycles as

1. -cycle→Loop
2. -cycle→Pair of parallel edges

3. -cycle→Triangle.
4. -cycle→Square,
5. -cycle→Pentagon.

The term circuit will always imply the presence of at least one distinct edge. That is, It is a path of length ≥ 1, (i.e., a closed trail of length three or more), with no repeated edges and whose end vertices are the same. The paths that originate and terminate at the same location is called a circuit. A circuit may have repeated vertices other than end vertices. A simple circuit is a closed walk that does not contain any repeated edges or repeated vertices except of course the first and last. A cycle is a circuit with no other repeated vertices except its end vertices if a vertex of a cycle is deleted then a path is obtained. This is not true for circuits, however.

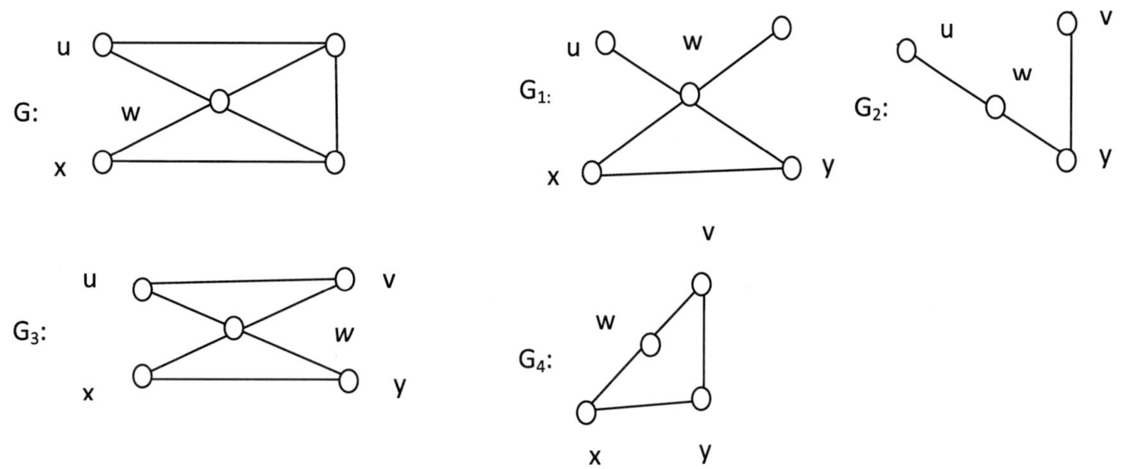

Figure 3.2 Trails, Paths, circuits, cycles as sub graphs of a graph

Furthermore, ***every cycle is a circuit and every circuit with at least two edges is a closed walk.*** Returning back to Figure 3.2, we have C : y, w, u, v, w, x, y; orC : x, y, w, u, v, w, x; or C : w, x, y, w, u, v, wis a circuit (not a cycle) while C′: x, y, v, w, x is a cycle, namely a 4-cycle.

The terms walk, open and closed walks, trail, path, circuit, and cycle for graphs have natural counterparts in digraphs as well. The difference being that the direction of the arc must be followed in each of these walks. The number of occurrences of arcs in the walk is its length. Digraph in which every vertex has positive out degree must contain cycles.

3.2 CONNECTEDNESS IN GRAPHS

A graph G is called **connected** if every two of its vertices are connected (i.e., iff there is a path between every pair of vertices) . That is, a graph G is connected if given any vertices u and v, it is possible to travel from u to v along a sequence of adjacent edges of the graph. In other words, in an undirected graph, if any pair of vertices is reachable from one another, then the graph G is connected.

An example of a connected graph is shown in Figure 3.3(a) and a disconnected graph with 2 components is shown in Figure 3.3(b).

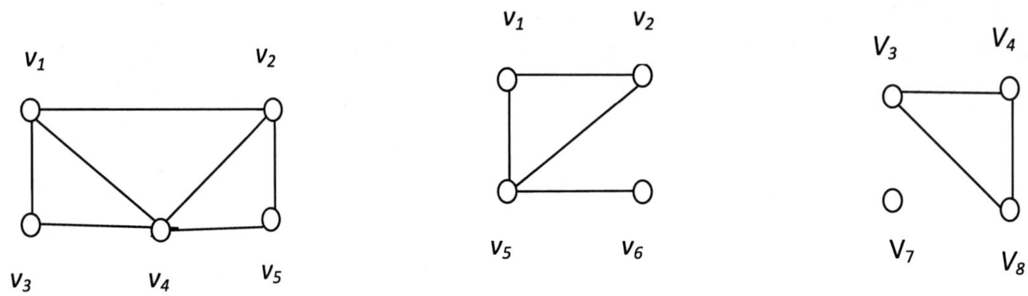

(a) A Connected graph **(b)** Disconnected graph with two components

Figure 3.3

The graph G_1 in Figure 3.3(a) is connected, because for every pair of distinct vertices there is a path between them. However, the graph G_2 in Figure 3.3(b) is not connected. For instance there is no path between vertices v_2 and v_4.

A disconnected graph is the union of two or more connected sub graphs each of which has no vertex in common. A disjoint sub graph H that is also called connected component of the graph G should satisfy following conditions:

i) H is a sub graph of G.

ii) No connected sub graph of G has H as sub graph and contains vertices or edges that are not in H. For example, the connected components of the graph shown in Figure 3.3(b) are $\{v_1, v_2, v_5, v_6\}$, $\{v_7\}$, $\{v_3, v_4, v_8\}$.

A Graph G is said to be ***Strongly connected,*** if for any pair of vertices u and v, there exists a path from u to v and vice versa. The above definition cannot be applied to directed graphs without some further modifications, because in a directed graph if a vertex is reachable from another vertex *v* the vertex v may not be reachable from *u*.

The connectedness in a directed graph G *(V, E)*, based on three notions, depending on whether the directions of the edges are considered, and is defined as follows:

i) A directed graph G is called connected or ***weakly connected*** if it is connected as an undirected graph where each directed edge is considered as an undirected edge (Figure 3.4(a)). That is, a digraph is weakly connected if and only if there is always a path between two vertices when the directions of the edges are disregarded.

ii) A directed graph G is called ***unilaterally connected graph*** if for any pair of vertices of the graph, at least one of these is reachable from other vertex. (Figure 3.4(b)).

iii) A directed graph G is called ***strongly connected graph*** if for any pair of vertices of the graph both of these are reachable from other vertex. (Figure 3.4(c)). Clearly, any strongly connected digraph is also weakly connected. That is, for the digraph to be strongly connected there must be a sequence of directed edges from any vertex in the graph to any other vertex. The sub graphs of a directed graph G that are strongly connected but not contained in larger strongly connected sub graphs, that is, the maximal strongly connected sub graphs, are called the strongly connected components or strong components of G. Unless otherwise stated connectedness implies connectedness in an undirected graph.

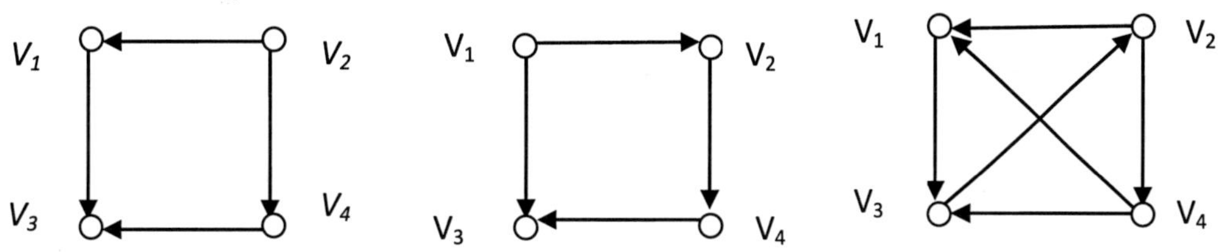

(a) Weakly Connected **(b)** Unilaterally Connected and **(c)** Strongly Connected Graph

Figure 3.4 Connected Digraphs

Example 1

For the graphs G and H shown in Figure 3.5 verify i) whether G and H are strongly connected? ii) whether G and H are weakly connected?

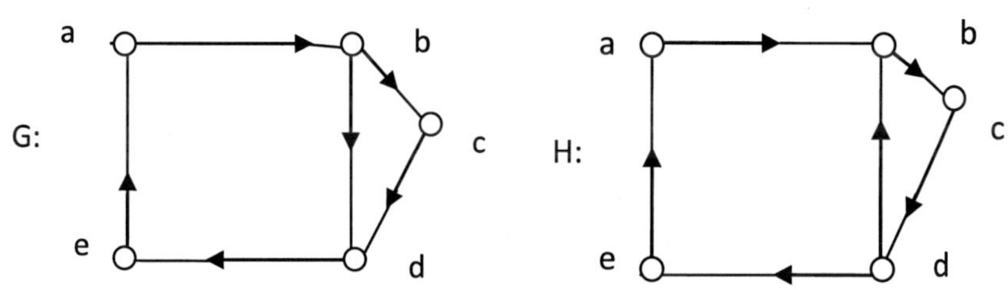

Figure 3.5 Directed Graphs G and H

Solution: G is strongly connected because there is a path between any two vertices. Hence G is also weakly connected. The graph H is not strongly connected. There is no directed path from a to b in this graph. However, H is weakly connected, because there is a path between any two vertices in the underlying graph of H.

For a simple digraph, the maximal strongly connected sub graph is called a strong component. Similarly, a maximal unilaterally connected or maximal weakly connected sub graph is called a unilateral or weak component respectively.

Example 2:

For the digraph H in Figure 3. 5 identify the strongly connected components?

Solution: The graph H in figure 3. 5 has three strongly connected components, consisting of the vertex a; the vertex e; and the graph consisting of the vertices b, c, and d and edges (b, c), (c, d), and (d, b).

An important characterization of connected in digraphs is depicted by the following theorem.

THEOREM 3.1

In a simple digraph G (V, E), every vertex of the digraph lies exactly in one strong component.

Proof: Let $v \in V$ and S be the set of all vertices of G which are mutually reachable with *v*. The set S naturally contains *v* and is a strong component of G. This shows that every vertex of G is contained in a strong component.

Now, assume that a vertex *v* is in two strong components. It implies that any vertex in one strong component which contains *v* is reachable from any vertex in the other strong component which also contains *v*, because every such path is easily established through *v*. This however is not possible. Hence every vertex is contained in exactly one strong component. Thus the strong components partition V.

Next, we may give some interesting and useful results involving connectedness of graphs.

THEOREM 3.2

In a graph G, every u-v path contains a simple u-v path

Proof*:* We proceed by induction on the length n of path P. If a path is closed, then it certainly contains the trivial path. Assume that P is a u-v path. If P is a path of length one, then P is itself a simple path. Assume that all open u-v paths of length K, where $1 \leq K \leq n$ contains a simple u-v path. More over, assume that P is the open u-v path $\{ v_0, v_1\}, \ldots, \{v_n, v_{n+1}\}$ where $u = v_0$ and $v = v_{n+1}$. Of course it may be that P has repeated but if not, then P is simple u-v path and we are done. If on the other hand, if P contains repeated vertices, and let i and j be distinct positive integers where $i < j$ and $v_i = v_j$. If the closed path v_i-v_j is removed from P, an open path P' is obtained having length $\leq n$ since at least one edge $\{v_i, v_{I+!})$ was deleted from p. By the induction hypotheses, P' contains a simple u-v path and so P as desired. Hence the theorem is proved.

THEOREM 3.3

A graph G is connected if and only if its vertex set V can be partitioned into two non empty, disjoint subsets V_1 and V_2 such that there exists no edge in G whose one end vertex is in subset V_1 and the other in subset V_2.

Proof: Suppose that such a partitioning exists. Consider two arbitrary vertices u and v of G, such that $u \in V_1$ and $v \in V_2$. No path can exist between vertices u and v; otherwise, there would be at least one edge whose one end vertex would be in V_1 and the other in V_2. Hence if a partition exists, G is not connected.

For the converse, assume that G is a disconnected graph. Let u be a vertex in G . Let V_1 be the set of all vertices that are joined by a path to u. Since G is disconnected, V_1 does not include all vertices of G. The remaining vertices will form a (non empty) set V_2. No vertex in V_1 is joined to any in V_2 by an edge. Hence the partition.

THEOREM 3.4

If a graph (connected or disconnected) has exactly two vertices of odd degree, there must be a path joining these two vertices

Proof: Let G be a graph with all even vertices except vertices v_1 and v_2, which are odd. Theorem 1.1, which holds for every graph and therefore for every component of a disconnected graph, says that, no graph can have odd number of odd vertices. Therefore, in graph G, v_1 and v_2 must belong to the same component, and hence must have a path between them.

THEOREM 3.5

A Simple graph (i.e., a graph without parallel edges or self loops) with n vertices and k components can have (n-k) (n-k+1) / 2 edges.

Proof: Suppose n_i represents the number of vertices in a component i, for $1 \leq i \leq k$ of a graph G, then the sum of all vertices in each of k components must equal the number of vertices in G, i.e. $\sum_{i=1}^{k} n_I = n$. Now the maximum possible number of edges in the ith componentof a simple connected graph with n_i vertices is = $\frac{1}{2} n_i(n_i - 1)$ when it is complete. Hence, the maximum number of edges in G are

$$\frac{1}{2}\sum_{i=1}^{k} n_I (n_i - 1) = \frac{1}{2}\sum_{i=1}^{k} n_i^2 - \frac{1}{2}\sum_{1}^{k} n_i$$

The proof of the theorem depends on the algebraic inequality,

$$\sum_{i=1}^{k} n_i^2 \leq n^2 - (k-1)(2n-k)$$

$$\frac{1}{2}\left[n^2 - (k-1)(2n\text{-}k)\right] - \frac{1}{2}n$$

$$\frac{1}{2}\left[n^2 - 2nk + k^2 + n\text{-}k\right] = \frac{1}{2}(n\text{-}k)(n\text{-}k+1)$$

THEOREM 3.6

If G is a graph of order n, If deg (u) + deg (v) $\geq$ n-1. For every non adjacent vertices u and v of G, then G is connected and diameter (G) $\leq$ 2.

Proof: The proof of the theorem is easy. We show that every two distinct vertices of G are connected by a path of length at most 2. Let x, y $\in$ V(G). If xy $\in$ E(G), then xy is a path and x and y are certainly connected. Hence we may assume that xy $\notin$ E(G). Therefore, deg(x) + deg(y) which implies that there must be a vertex w that is adjaceant to both x and y. So x, w, y is a path in G. asdesired. Theorem 3.6 implies that if G is a graph of order n such that deg(v) $\geq \frac{n-1}{2}$ for every vertex v of G, then G must be connected.

THEOREM 3.7

If G is a graph of order n with $\delta(G) \geq \frac{n-1}{2}$, then G is connected.

Proof: For every nonadjacent vertices u and v of $G \deg(u) + \deg(v) \geq \frac{n-1}{2} + \frac{n-1}{2} = n$

1. By theorem 2.5, G is connected.

THEOREM 3.8

If a simple graph G is not connected, then G^c is connected.

Proof: Let u and v be any two vertices ofG^c (and therefore of G). If u and v belong to different components of G, then obviously u and v are non adjacent in G and so they are adjacent in G^c. Thus u and v are connected in G^c. In case u and v belong to the same component of G, take a vertex w of G not belonging to this component of G. Then uw and vw are not edges of G and hence they are edges of G^c. Then uwv is a u-v path in G^c. Thus G^c is connected as desired.

THEOREM 3.9

If a simple graph G with n vertices has more than $\frac{1}{2}(n-1)(n-2)$ edges, then G is connected.

Proof: Let G be a simple graph with n vertices and more than $\frac{1}{2}(n-1)(n-2)$ edges. If G has k components, then by heorem 2.5, G can have at most $\frac{1}{2}(n-k)(n-k+2)$ edges. But G has more than $\frac{1}{2}(n-1)(n-2)$ edges. Therefore, $\frac{1}{2}(n-k)(n-k+2) > \frac{1}{2}(n-1)(n-2)$ which is possible onlywhen k = 1, Hence G is connected.

3.3 EULER GRAPHS

In the earlier section you have read that a trail in a graph G is a walk in G in which the edges are distinct, or in other words, no edge of G appears in the trail more than once. Now, we define a trail as ***Euler trail*** if it includes every edge of G. So, A trail is Euler if each edge of G is in the trail exactly once. In general, if there is an open trail from vertex *u* to *v* in G and this trail traverses each edge in G exactly once, the trail is called ***Euler trail.***

On the other hand, A ***tour*** of G is a closed walk of G which includes every edge of G at least once. An ***Euler tour*** of G is a tour which includes each edge of G exactly once. So, an ***Euler tour*** is just a closed Euler trail. A graph G is called ***Eulerian*** or Euler if it has an ***Euler tour*** i.e., if some closed walk in a graph contain all the edges of the graph, then the walk is called an Euler line and the graph an Euler graph. For instance the Graph in Figure 3.6 is ***Eulerian***. By its very definition a walk is always connected. Since the Euler line (which is a walk) contains all edges of the graph, an Euler graph is always connected, except for any isolated vertices the graph may have. Since the isolated vertices do not contribute anything to the understanding of an Euler graph, it is assumed that Euler graphs do not have any isolated vertices and are therefore connected. More formally, an Eulertour of G, if it exists can be described from any vertex of G, a clear indication that every Eulerian graph is connected.

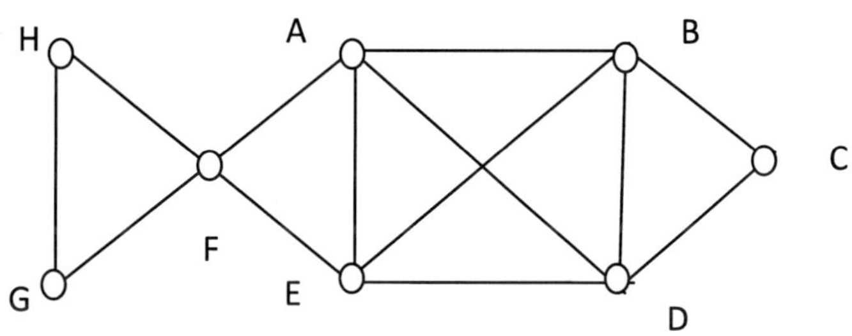

Figure 3.6

Since any closed walk in an undirected graph enters and leaves any vertex the same number of times the sub graph composed of the edges in any closed walk is even. Conversely, if the graph is even, then it contains an Euler tour.

A path that passes through each edge exactly once, but vertices may be repeated, is called Euler path. If the path is a circuit, then it is called an Euler circuit. A graph that contains an Euler tour (path or circuit) is called an Eulerian graph. For instance, in the graph of Figure 3.7, the graph G_1 has an Euler circuit, namely, a, e, c, d, e, b. a. Neither of the graphs G_2 or G_3has an Euler circuit. However, G_3 has an Euler path namely, a, c, d, e, b, d, a, b. G_2 does not have an Euler path.

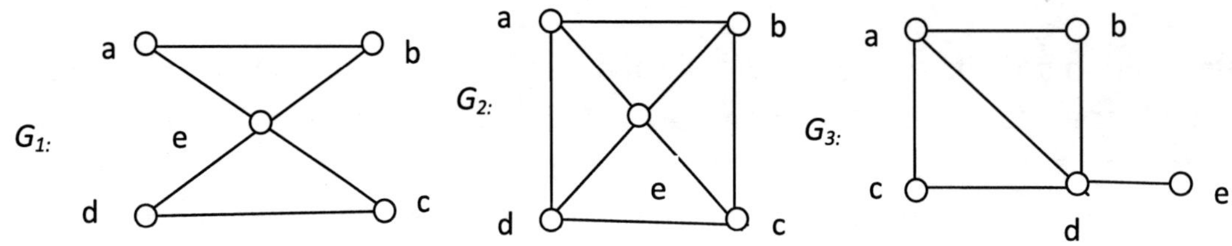

Figure 3.7 The undirected graphs G_1, G_2 and G_3.

A directed graph that contain an ***Eulerian circuit*** is called ***Eulerian Digraph*** (Fig 3.8).

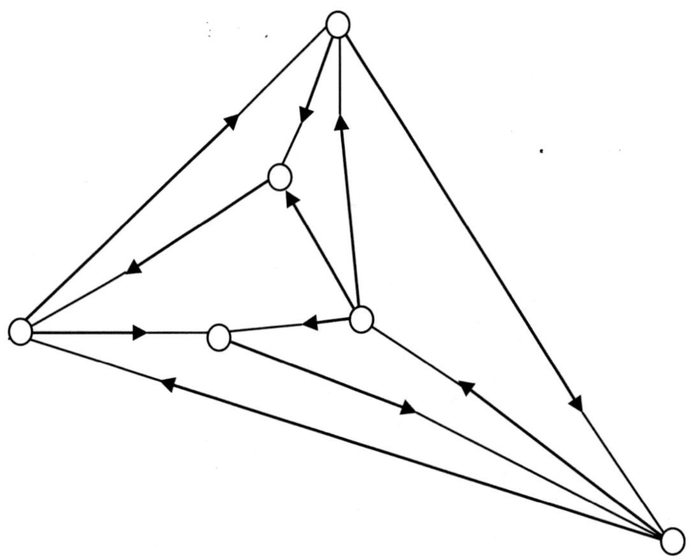

Figure 3.8 Eulerian Digraph

We will now prove the main theorem (given in Euler's 1736 paper) characterizing undirected multi graphs that have Euler path.

Theorem 3.10

A Given connected graph is an Euler graph if and only if all vertices of G are of even degree.

Proof: Let the Multigraph G is an Euler graph. It therefore contains an Euler line (which is a closed walk). It is clear that G must be connected. In tracing this walk we observe that every time the walk meets a vertex v it goes through two "new "edges incident on v- with one we 'entered' v and with the other "exited. "This is true not only of all intermediate vertices of the walk but also of the terminal vertex, because we "exited" and "entered" the same vertex at the beginning and end of walk, respectively. Thus if G is an Euler Graph, the degree of every vertex is even.

Konisberg Bridge Problem: Now looking at the Graph of Konigsberg Bridges (Figure 1.1), we find that not all of its vertices are of even degree. Hence it is not an Euler Graph. Thus it is not possible to walk over each of these seven bridges exactly once and return to the starting point. The proofs of the following corollaries follow from the preceding Theorem 3.10.

Corollary 1

A directed multigraph G has an Eulerian path iff it is unilaterally connected and the in degree of each vertex is equal to its out degree, with possible exception of two vertices, for which it may be that the in degree one is larger than its out degree and the in degree of the other is one less than its out degree.

Corollary 2

A directed Multigraph G has an Eulerian circuit Iff the in degree of every vertex in G is equal to its out degree.

Example 3:

Verify whether the graphs G_1, G_2, and G_3 in Figure 2.9 has Euler path / Euler circuit?

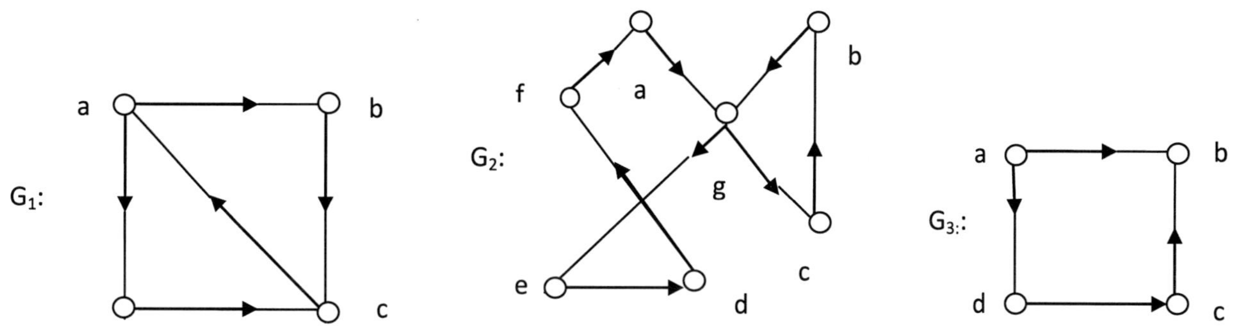

Figure 3.9

Solution

The graph G_1 is unilaterally connected and clearly the degrees of the vertices are satisfying the corollary on Euler path for directed multi graphs. Therefore it has Euler path: c-a-b-c-d-b.

V(G)	In degree	out degree
a	1	1
b	2	1
c	1	2
d	1	1

As the degrees of the vertices are not satisfying the corollary on Euler circuit therefore the graph has no Euler circuit. Therefore, the given graph has only Euler Path.

(ii) For the Graph G_2(which is unilaterally connected) in Figure 3.9, the degrees of all vertices are satisfying the corollary on Euler circuit for multi graphs. Therefore the given graph has Euler circuit: a-g-c-b-g-e-d-f-a.

V(G)	In degree	Out degree
A	1	1
B	1	1
C	1	1
D	1	1
E	1	1
F	2	2

(iii) In Graph G_3 of Figure 3.9, d is not reachable from b. That is, the given graph is not unilaterally connected. Therefore, the given graph has no Euler path and no Euler circuit.

There are situations, where one may ask about a walk, not necessarily closed, that visits each edge precisely once. Since the number of odd degree vertices was used to characterize Euler circuits, a nice approach would be to consider the number. In the original paper of Euler's, he pointed out another classical result in graph theory that relates to number is given below.

THEOREM 3.11

A Graph G with no isolated vertices has an Eulerian trail if and only if G is connected and has exactly two vertices of odd degree.

Proof: Suppose G is connected and u and v are vertices of G with odd degrees. Add an additional edge e' = {u, v} to G, so that the new graph G_1 so obtained is connected and all vertices in G_1 have even degree. Hence G, is Euler and so has Eulerian trail.

For the converse, assume that G has an Eulerian trail which is not closed. Since each vertex is associated with two edges and also there is only one edge associated with each end vertex of the trail, all end vertices are of odd degree where as others are of even degree.

Eulerian Graphs admits among others, the following elegant characterizations, Theorems 3.10, 3.11, and 3.12.

THEOREM 3.12

A connected graph G is Euler iff G has cycles C_1...........C_n such that every edge of G belongs to exactly one cycle C_i i.e., G is the union of edge disjoint cycles.

Proof: To prove this, we need only to show that, if G can be decomposed into cycles, then G is union of edge disjoint cycles in G. Since the degree of every vertex in a cycle is two, the degree of every vertex in G is even. Hence G is Eulerian. Conversely, let G be Eulerian. As the degree of every vertex is even, minimum degree $\delta(G) \geq 2$. So G contains a cycle C_1. Again degree of every vertex is even in G-C_1. If G-C_1 has at least one edge, then it has a connected component in which minimum degree $\delta(G) \geq 2$. So G-C_1 contains a cycle $C_{2,}$ If G is not the union of C_1 and C_2 consider the graph G–(C_1 C_2). As degree of every vertex is even in G–(C_1 C_2). it contains a cycle C_3. Continue this process until no more edge is left. Hence, at the final step we obtain a decomposition of G into a union of edge disjoint cycles.

Note 1:

i) If a graph G has a vertex of odd degree, then there can be no Euler circuit in G.

ii) If a graph G is connected and each vertex has even degree, then there is an Euler circuit in G.

iii) If a graph G has more than two vertices of odd degree, then there can be no Euler path in G.

iv) If a graph G is connected and has exactly two vertices of odd degree, then there is an Euler path in G. Any Euler path in G must begin at one vertex of odd degree and end at the other.

We now present two algorithms that will help us to efficiently decide whether a given graph contain an Eulerian Circuit.

3.4 FLEURY'S ALGORITHM

Fluery's Algorithm is useful in constructing an Euler circuit in a connected graph G. This algorithm constructs an Euler tour in a Euler Graph. Its approach is to construct a trail that grows to the desired Eulerian tour. The Fleury's algorithm for constructing Euler circuits begins with an arbitrary vertex of connected multi graph and forms a circuit by choosing edges successively. Once an edge is chosen, it is removed. Edges are chosen successively, so that each edge begins where the last edge ends, and so that this edge is not a cut edge, unless there is no alternative.

Step 1: Choose any vertex v_0 in the Euler Graph. And set $W_0 = v_0$.

Step 2: If the trail $\mathbf{W_i = v_0, e_1, v_1 \ldots\ldots\ldots\ldots e_i, v_i}$ has been chosen, choose an edge e_{i+1} different from $e_1, \ldots\ldots\ldots\ldots\ldots e_i$ such that

(i) e_{i+1} is incident with v_i and

(ii) unless there is no alternative, e_{i+1} is not a bridge of the edge deleted sub graph G–($e_1, \ldots\ldots\ldots e_i$).

Step 3: Stop, if W_i contains every edge of G, otherwise repeat step 2.

Example 4: Use Fluery's algorithm on the below graph to find an Euler circuit?

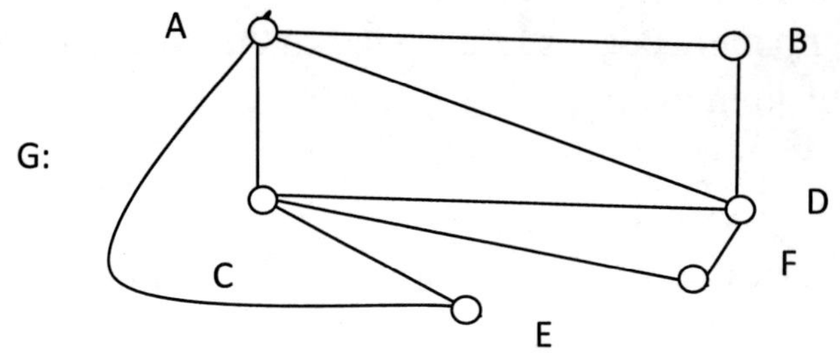

Figure 3.10 (a)

Solution

Choose the vertex F. The result of applying step 2 are listed in the following table (Figure 3.10(b)).

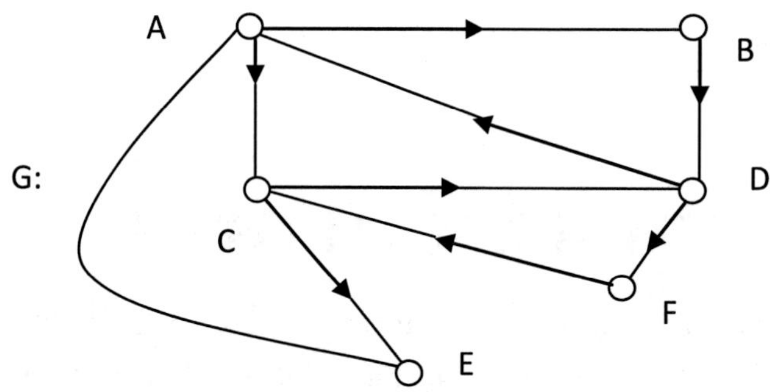

Figure 3.10 (b)

Current path	Next edge	Reasoning
F	FC	Travel from F to C. Could have also gone from F to D
FC	CD	Travel from C to D.
FCD	DA	Travel from D to A. Could have also gone to B, but not to F. DF is a bridge.
FCDA	AC	Travel from A to C. Could have also gone to E. But not to B. AB is a bridge.
FCDAC	CE	Travel from C to E . No alternative.
FCDACE	EA	Travel from E to A. No alternative.
FCDACEA	AB	Travel from A to B . No alternative.
FCDACEAB	BD	Travel from B to D. No alternative.
FCDACEABD	DF	Travel from D to F . No alternative.
FCDACEABDF	No edge	End. (No edge remaining).

3.5 HIERHOLZER'S ALGORITHM

Fluery's algorithm is a complex algorithm. There is another algorithm developed by Hierholzer. Its method is to start with any closed trail in G and "attach" to it, systematically, "detour" trail until all edges of G are used up.

Step 1: Choose any vertex v in G and choose any closed trail W_0 in G starting and ending at v. Set i=0.

Step 2: If E (W_i) = E (G) stop. Since then W_i is an Euler tour of G. Here, E (W_i) denotes the set of edges of a sub graph $W_{i.}$ Otherwise choose a vertex v_i on W_i. which is incident with an edge in G which is not in W_i. Now, choose a closed trail W_i^* in the sub graph G-E (W_i.), Starting at the vertex v_i . ($W_i^*$is a detourtrail).

Step 3: Let W_{i+1} be the closed trail consisting of the edges of both W_i and W_i^* obtained by starting at the vertex v, traversing the trail $W_i$until v_i is reached, then traversing the closed trail $W_i^*$and, on returning to v_i, completing the rest of the trail $W_{i..}$ Set i= i+1 and return to step 2.

Example 5:

Using **Hierholzer Algorithm,** produce a Euler tour for he graph given below starting with $w_0 = a_2\, a_4\, a_7\, a_2$. ?.

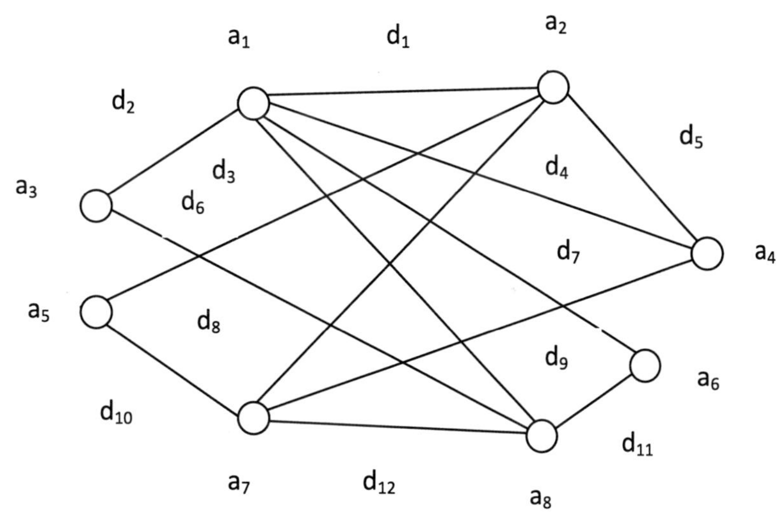

Figure 3.11

Solution

Iteration 1:

Step 1: $W_0 = a_2\, a_4\, a_7\, a_2$ or $a_2\, d_5\, a_4\, d_9\, a_7\, d_4\, a_2$, $i = 0$.

Step 2: Choose a_7 on W_0 incident with d_{12} not in W_i .

Choose $W_0^* = a_7\, d_{12}\, a_8\, d_8 a_3 d_2\, a_1\, d_1\, a_2\, d_6\, a_5\, d_{10}\, a_7$.

Note that all d_i G-E (W_0).

Step 3: Let $W_{0+1} = W_1 = \underbrace{a_2\, a_5\, a_4\, d_9\, a_7}_{W_0}\ \underbrace{d_{12}\, a_8}_{W_1}\, d_8\, d_3\, a_3\, d_2\, a_1\, d_1\, a_2\, d_6\, a_5\, d_{10}\, a_7\ \underbrace{d_4\, a_2}_{W_0}.$

Set i =0 +1 =1, Go to step 2.

Iteration 2:

Step 1: Choose a_8 on W_1 incident with d_{11} not in W_1.

Choose $W_1^* = a_8\, d_{11}\, a_6\, d_7\, a_1\, d_3\, a_8$.

Note that all d_i G-E (W_1).

Set i =2, Go to step 2.

Step 2: Here, E (W_2) = E (G). Hence stop. W_i is an Euler tour of G.

Step 3: Let $W_2 = W_{1+1} = \underbrace{a_2 a_5 a_4\, d_9 a_7 d_{12}}_{W_1}\ \ \underbrace{a_8 d_{11} a_6 d_7 a_1 d_3}_{W_1^*}\ \ \underbrace{a_8 d_8 a_3 d_2 a_1 d_1 a_2 d_6 a_5 d_{10} a_7 d_4 a_2}_{\text{Remaining } W_1}.$

3.6 HAMILTONIAN GRAPHS

Hamiltonian graphs are named after Sir William Hamilton, an Irish mathematician, who invented a puzzle called, Icosian Game. The puzzle involved a dodecahedron on which each of the 20 vertices was labeled by the name of the some city . The object of the game was to construct, using the edges of dodecahedron, a tour of all cities which visited each city exactly once, beginning and ending at the same city. The graph of this puzzle is **Hamiltonian graph**.

A **Hamiltonian path** in a graph G is a path which contains every vertex of G. i.e., it contains every vertex of G once and only once. A Hamiltonian Cycle G is a cycle which contains every vertex of G . Now, A graph G is called **Hamiltonian**. If it has a **Hamiltonian cycle**. For example, The Graph $K_{3,3}$ is a Hamiltonian graph While the Petersen graph (Q_3) is non Hamiltonian. The complete bipartite graph $K_{m,n}$ is Hamiltonian if and only if m = n and n > 1.

Obviously, if a graph contains a Hamiltonian cycle, then it contains a Hamiltonian Path. If a graph contains a Hamiltonian path, it need not contain a Hamiltonian cycle. For Example, the graph P_n (path of length n) clearly contains a Hamiltonian path, But P_n contains no cycle at all.

In considering the existence of a Hamiltonian cycle (or path) we need only consider simple graphs. This is because, the Hamiltonian circuit (path) traverses every vertex exactly once. Hence it cannot include a self loop or a set of parallel edges. Thus, a general graph may be made simple by removing parallel edges and self loops before looking for a Hamiltonian circuit in it. A graph is Hamiltonian its underlying simple graph is Hamiltonian.

A simple graph G is called **Maximal Hamiltonian.** If it is not Hamiltonian but the addition of any edge connecting two non-adjacent verticesforms a Hamiltonian graph.

The following observations can be made about Hamiltonian graphs:

1. Euler circuit is a circuit that traverses each edge exactly once and, therefore each vertex at least once. A Hamiltonian cycle traverses each vertex exactly once(and hence may miss some edges altogether). Thus, there is a striking similarity between the concepts of Eulerian graph and Hamiltonian graph. But till now there is no characterization of Hamiltonian graphs as in the case of Eulerian graphs.

2. A Hamiltonian cycle always provides a Hamiltonian path, by deleting any edge. On the other hand, a Hamiltonian path may not lead to a Hamiltonian cycle (it depends on whether or not the end points of the path happen to be joined by an edge in the graph)

3. The problem of proving that no Hamiltonian cycle (or path) exists in a given graph is very difficult.

4. To prove non existence of a Hamiltonian path or cycle, we must begin building parts of a Hamiltonian path and show that the construction always fail, that is, we cannot visit all .

5. The idea underlying is that a Hamiltonian cycle must contain exactly two edges incident at each vertex and a Hamiltonian path must contain at least one of the edges.

Note 2*:* Unlike the situation for Euler cycles no easily verified necessary and sufficient conditions are known for the existence of a Hamiltonian cycle in a graph.

We list here some basic rules for constructing Hamiltonian paths and cycles.

Rule 1: If G has n vertices, then a Hamiltonian path (if it exists) must contain exactly n-1 edges and a Hamiltonian Cycle must contain exactly n edges.

Rule 2: If a vertex v in G has degree k, then a Hamiltonian path must contain at least one edge incident on v and at most two edges incident on v. A Hamiltonian cycle contains exactly two edges incident on v. In particular, both edges incident on a vertex of degree two will be contained in every Hamiltonian cycle. In sum: there cannot be three or more edges incident with one vertex in a Hamiltonian cycle.

Rule 3: No cycle that does not contain all the vertices of G can be formed when building a Hamiltonian path or cycle.

Rule 4: Once the Hamiltonian cycle we are building has passed through a vertex v, then all other unused edges incident on v can be deleted because only two edges incident on v can be included in a Hamiltonian cycle.

A given graph may contain more than one Hamiltonian circuit. Of interest, are all the edge disjoint Hamiltonian circuits in a graph, the determination of the exact number of which in a graph in general is also an unsolved problem. The Hamiltonian circuit in a complete graph with odd number of vertices is given by Theorem 3.11.

THEOREM 3.13 (ORE'S THEOREM)

Let G be simple graph with $n \geq 3$ vertices. If for every pair of non adjacent vertices u, v of G, $\deg(u) + \deg(v) \geq n$, then G is **Hamiltonian.**

Proof: Suppose that G satisfies the condition of the theorem. But G is not Hamiltonian. Add edges to G (without adding vertices) and get a super graph G^* of G such that G^* is a maximal simple graph that satisfies the condition of the theorem but G^* is non-Hamiltonian. Such a graph G^* must exist since G is non-Hamiltonian while the complete graphon V(G) is Hamiltonian. Hence, for any pair u and v of non adjacent vertices of G^*, G^*+uv must contain a Hamiltonian cycle C. This cycle C would certainly contain the edge e = uv. Then C-e is a Hamiltonian path $u = v_1v_2....v_n = v$ of G^*.

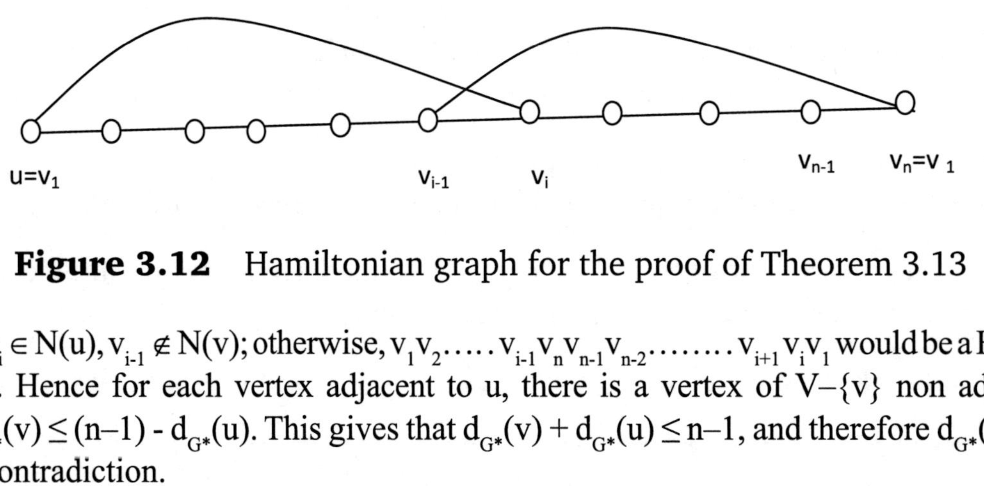

Figure 3.12 Hamiltonian graph for the proof of Theorem 3.13

Now if $v_i \in N(u)$, $v_{i-1} \notin N(v)$; otherwise, $v_1v_2....v_{i-1}v_nv_{n-1}v_{n-2}.......v_{i+1}v_iv_1$ would be a Hamiltonian cycle in G^*. Hence for each vertex adjacent to u, there is a vertex of $V-\{v\}$ non adjacent to v. But then $d_{G*}(v) \leq (n-1) - d_{G*}(u)$. This gives that $d_{G*}(v) + d_{G*}(u) \leq n-1$, and therefore $d_{G*}(v) + d_{G*}(u)) \leq (n-1)$, a contradiction.

It would seem that a graph with many edges should have a good chance of being Hamiltonian. The following theorem published by Dirac in 1952, provides a further evidence for the support of this idea.

Corollary 3 (Dirac's Theorem)

The connected graph G with $|V| \geq n$ vertices has a Hamiltonian circuit if $\deg(v_I) \geq n/2$, for each vertex n_i.

Proof: Dirac's theorem can be proved as a corollary to Ore's theorem, because the conditions of Dirac imply those of Ore's theorem. That is, deg(u) + deg(v) $\geq$ n for every pair of non adjacent vertices u and v.

Both Dirac' and Ore's theorem provide sufficient conditions for a connected simple graph to have a Hamiltonian circuit. However, these theorems do not provide necessary conditions for the existence of a Hamiltonian circuit. For example, the graph C_5 has a Hamiltonian circuit but does not satisfy the hypothesis's of either Ore's or Dirac's theorem.

THEOREM 3.14

Let u and v be non adjacent vertices in a graph G of order n such that deg (u) + deg(v) $\geq$ n, Then G + uv is Hamiltonian if and only if G is Hamiltonian.

Proof: If G is a Hamiltonian Graph, then certainly G + uv is Hamiltonian for any two non adjacent vertices u and v of G. Thus we need only to verify the converse.

Let G + uv be a Hamiltonian graph for two non adjacent vertices u and v of a graph G, and assume, to the contrary, that G is not Hamiltonian. This implies that every Hamiltonian cycle in G + uv must contain the edge uv and so G contains a Hamiltonian u-v path. Since $\deg_G$ u + $\deg_G$ v $\geq$ n, G contains a Hamiltonian cycle. This is a contradiction.

3.7 CLOSURE OF A GRAPH

Let G be a simple graph. If there are two non adjacent vertices u_1 and v_1 in G such that $d(u_1)$+ $d(v_1) \geq n$ in G . Join u_1 and v_1 by an edge to form the super graph G_1. Then, if two vertices u_2 and v_2 are such that $d(u_2) + d(v_2) \geq n$ in G_1, join u_2 and v_2 by an edge to form the super graph G_2. Continue in this way to join pairs of non adjacent vertices whose degree sum is at least n. The final super graph thus obtained is called closure of G and is denoted by C(G). That is, the closure C(G) of a graph G of order n is the graph obtained from G by joining pairs of nonadjacent vertices whose degree sum is at least n (in the resulting graph at each stage) until no such pair remains. Figure 3.11 shows a graph and its closure.

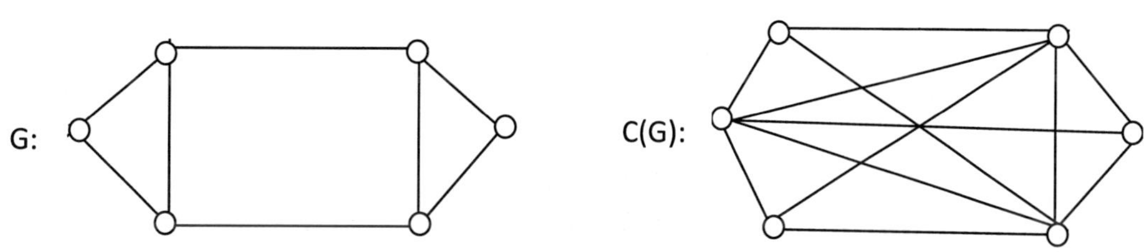

Figure 3.13 A graph G and its closure C(G).

There is a Theorem due to Bondy and Chva'tal which is as follows:

THEOREM 3.15

Let G be a Graph of order $V_G \geq 3$. (i) G is Hamiltonian if and only if its Closure C(G) is Hamiltonian. (ii) If C(G) is a complete Graph, then G is Hamiltonian. .

Proof: First, $G \subseteq C(G)$ and G spans C(G), and thus if G is Hamiltonian so is C(G).

In the other direction, Let $G = G_0, G_1, \ldots\ldots, G_k = C(G)$ be a construction sequence of the C(G). If C(G) is Hamiltonian, then so are $G_{k-1}, \ldots\ldots, G_1$ and G_0 by Theorem. The claim (ii) follows from (i) since each complete graph is Hamiltonian.

3.8 TRAVELLING–SALESMAN PROBLEM

Hamiltonian paths and circuits can be used to solve practical problems. For example, many practical applications ask for a path or circuit that visits each road intersections in a city or each node in a net work exactly once. Finding a Hamiltonian path or circuit in the appropriate graph model can solve such problems. The famous travelling salesman problem asks forthe shortest route a sales man should take to visit a set of cities.

In Graph theory terms, we have a weighted graph with vertices (towns) joined by a weighted edge (iff road connects them). This graph has numerous Hamiltonian Circuits and we are to pick one that has the smallest sum of distances (or weights). Theoretically, the problem of the travelling salesman can always be solved by enumerating all (n-1)|/2 Hamiltonian circuits, calculating the distance travelled in each, and then picking the shortest one. However, fora large value of n, the labor involved is too great even for a digital computer The problem is to prescribe a manageable algorithm for finding the shortest route. No efficient algorithm for problems of arbitrary size yet has been found, although many attempts have been made. There are also available several heuristic methods of solution that give very close to the shortest one, but do not guarantee the shortest. We will restrict ourselves to complete weighted graphs and give two algorithms which give "reasonably good" solutions.

3.8.1 TWO OPTIMAL ALGORITHM

In this Algorithm we choose a Hamiltonian cycle C of the complete weighted Graph G and perform a sequence of modifications to C to find a cycle of smaller weight. Given below are the steps to perform this algorithm.

Step 1: Let $C = v_1v_2. \ldots\ldots\ldots\ldots v_nv_1$ be any Hamiltonian cycle of the weighted

graph G and letw be the weight of C, i. e.

$w = w(v_1v_2) + \ldots\ldots\ldots\ldots\ldots\ldots\ldots w(v_nv_1)$

Step 2: Set $j = i + 2$

Step 3: Set j=i+2

Step 4: Let C_{ij} be the Hamiltonian cycle

$C_{ij} = v_1v_2........v_i\ v_j\ v_{j-1}.......\ v_{i+1}v_{j+1}\ v_{j+2}..........\ v_nv_1$ and let w_{Ij}

Denote the weight of C_{ij} so that,

$W_{ij} = w\text{-}w\ (v_i, v_{i+1}) - w\ (v_j\ v_{j+1}) + w(v_i\ v_j)+w(v_{i+1}\ v_{j+1})$

If $W_{ij} < W$ then replace C by C_{ij} and go to step 1.

Step 5: Set j=j+1, if $j \leq m \rightarrow$ Go to step 4.

Otherwise,

Set i = i+1, if $i \leq n\text{-}2 \rightarrow$ Go to step 3

Otherwise, stop.

3.8.2 THE CLOSEST INSERTION ALGORITHM

The Closest Insertion Algorithm gradually build up a sequence of cycles in the graph which involve more and more vertices, until all the vertices are used up, and to involve one more vertex at each stage by determining which vertex is nearest to the cycle already created and then inserting this vertex into the cycle in cheapest possible way. The following steps are required.

Step 1: Choose any vertex v_1 as a starting vertex.

Step 2: From among the n-1 vertices not chosen so far, find one, say v_2, Which is closest to v_1. Let W_2 denote the walk $v_1v_2v_1$.

Step 3: From among the n-2 vertices not chosen so far, find one, say v_3 which is closest to the walk $W_2 = v_1v_2v_1$. . Let W_3 be the walk $v_1v_2v_3v_1$.

Step 4: From among the n-3 vertices not chosen so far, find one, say V_4, which is closest to the walk W_3. Determine which of the walks $v_1v_2v_3v_4v_1$, $v_1v_2v_4v_3v_1$, $v_1v_4v_2v_3v_1$ Is the shortest.

Let W_4 denote the shortest one and relabel it, if necessary, as $v_1v_2v_3v_4v_1$.

Step 5: From among the n-4 vertices not chosen so far, find one, say v_5 Which is closest to W_4. Determine which of the cycles $v_1v_2v_3v_4v_5v_1$, $v_1v_2v_3v_5v_4v_1$, $v_1v_2v_5v_3v_4v_1$, $v_1v_5v_2v_3v_4v_1$ is shortest.

Let W_5 denote the shortest one and relabel it, if necessary as, v1v2v3v4v5v1. Continue till all the vertex are chosen. Wngiven a Hamiltonian cycle of G which is approximately optimal.

Example 6: Carry out two optimal method and the closest insertion method for the travelling salesman problem for the complete weighted graph.

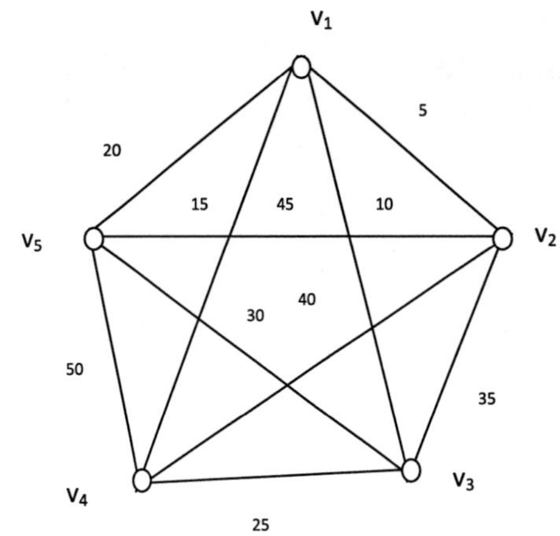

Figure 3.14

Solution: Two optimal Method

Step 1: Let C be $v_1v_2v_3v_4v_5v_1$. So that, w = 5+35+25+50+20 = 135

Step 2: Set i = 1

Step 3: Set j = i + 2 =1 + 2 = 3

Step 4: Set $C_{13} = v_1v_3v_2v_4v_5v_1$

So that, W_{13} = 10+35+40+50+20 = 155

Here $W_{13} >$ w.

Step 5: Increase j = 3 + 1 = 4

Step 4: Set $C_{14} = v_1v_4v_3v_2v_5v_1$

W_{14} = 15+25+35+45+20 = 140

Here $W_{12} >$ w.

Step 5: Increase j = 4 + 1 = 5

Step 4: Set $C_5 = v_1v_5v_4v_3v_2v_1$

W_{15} = 20+50+25+35+5 = 135

This is just reverse of the cycle C. Since W_{15} = w, we move on to step 2.

Step 2: i = i + 1 = 2

Since i ≤ 5-2 = n-2, we go to step 3

Step 3: j = i + 2 = 2 + 2 = 4.

Step 4: Set $C_{24} = v_1v_2v_4v_3v_5v_1$

$W_{24} = 5+40+25+50+20 = 140.$

Since $W_{24} > w$, we move to step 2.

Step 5: $j = j + 1 = 5$

Step 4: Set $C_{25} = v_1v_2v_5v_4v_3v_1$

$W_{25} = 5+45+50+25+10 = 135$

Since $W_{25} = w$, we move to step 2.

Step 2: $i = i + 1 = 3$

Since $i \leq 5\text{-}2 = n\text{-}2$, we go to step 3.

Step 3: $j = i + 2 = 5$

Step 4: Set $C_{35} = v_1v_2v_3v_5v_4v_1$

$W_{35} = 5+35+30+50+15 = 135$

Since, $W_{35} = 2$, we move to step 2.

Step 2: $i = i +1 = 4$

Step 3: $j = i+2 = 6 \geq n$ and $i \geq n\text{-}2$. stop. The optimal tour is,

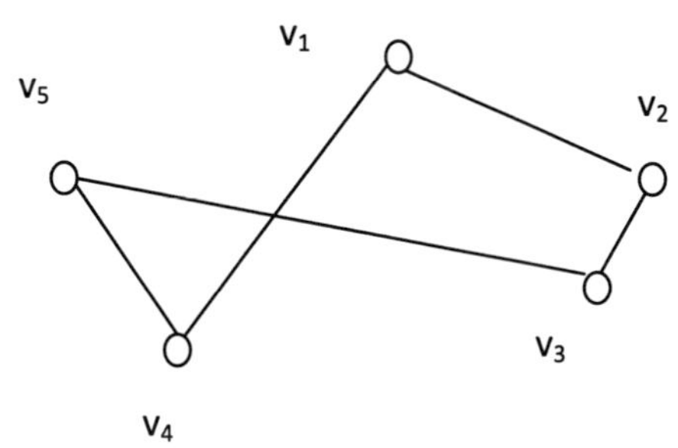

Figure 3.15

The Closest Insertion Method

Step 1: Choose v_1

Step 2: v_2 is closest to v_1 so $W_2 = v_1v_2v_1$

Step 3: v_3 is closest to W_2 (It has distance 10 from v_1). W_3 is the cycle $v_1v_2v_3v_1$

Step 4: v_4 is closest to W_3 (it has distance 15 from v_1). Find the length of Cycles

$v_1v_2v_3v_4v_1 = 5 + 35 + 25 + 15 = 80$

$v_1v_2v_4v_3v_1 = 5 + 40 + 25 + 10 = 80$

$v_1v_4v_2v_3v_1 = 15 + 40 + 35 + 10 = 100$

We can take any of first two, lets take $W_4 = v_1v_2v_3v_4v_1$

Step 5: v_5 is closest to W_4(it has distance 20 from v_1)

Find length of Cycles.

$v_1v_2v_3v_4v_5v_1 = 5 + 35 + 25 + 50 + 20 = 135$

$v_1v_2v_3v_5v_4v_1 = 5 + 35 + 30 + 50 + 15 = 135$

$v_1v_2v_5v_3v_4v_1 = 5 + 45 + 30 + 25 + 15 = 120$

$v_1v_5v_2v_3v_4v_1 = 20 + 45 + 35 + 25 + 15 = 140$

Cycle with length 120 is shortest, so

$W_5 = v_1v_2v_5v_3v_4v_1$ is shortest of these and hence the solution (Figure 3.16).

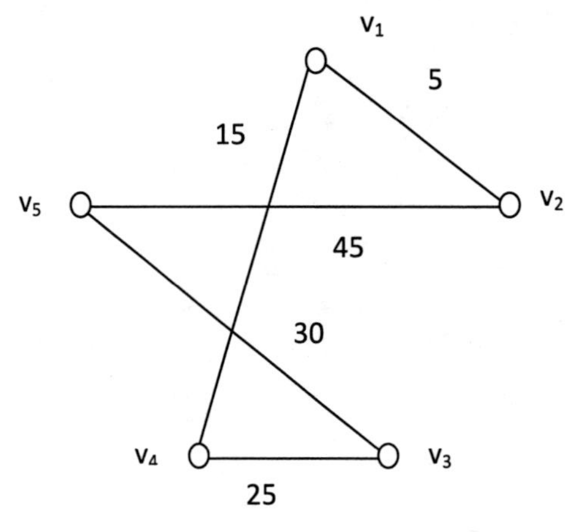

Figure 3.16

3.9 SHORTEST PATH PROBLEMS

In this section, we shall discuss two algorithms to determine the shortest distance between a designated vertex and each of the other vertices in a loop-free connected undirected graph.

3.9.1 Shortest Path in a Unweighted Graph

The shortest distance path in a loop-free unweighted graph G (V, E) is the path between two vertices v_i and v_j that uses the least number of edges. (called the shortest path). We shall discuss

Breadth First Search (BFS) and ***Back Tracking Search*** (BTS) algorithms to find the shortest path in an unweighted graph.

Steps of BFS Algorithm

1. Set the count i =0 and assign the first vertex v_0 = s (say).
2. Search all other unassigned adjacent vertices of vertex v_o = s in G. If there is no such vertex or vertices, then vertex v_0 is isolated, otherwise set the counter at i = i +1 = t and assign v_{i+1}= t and find the length of a shortest path from s to t.

Once the length of the shortest path is found from the previous algorithm, we use the Back tracking algorithm to find the actual shortest path from s to t. This algorithm uses the label λ(v) which are generated in the BFS algorithm.

Steps of BTS Algorithm

1. Set the counter at i= λ(t) and assign the last vertex v_i = t.
2. Search an unassigned adjacent vertex u to v_j and assign u = v_{i-1}, where λ(u) = i-1.
3. If i=1 stop, otherwise decrease I to i-2 and go to step 2.

In general there may be many shortest paths from sto t and the previous algorithm finds just one of them.

Example 7:

Find the shortest path from vertex v_1 to v_6 and its length in the graph shown in Figure 2.17.

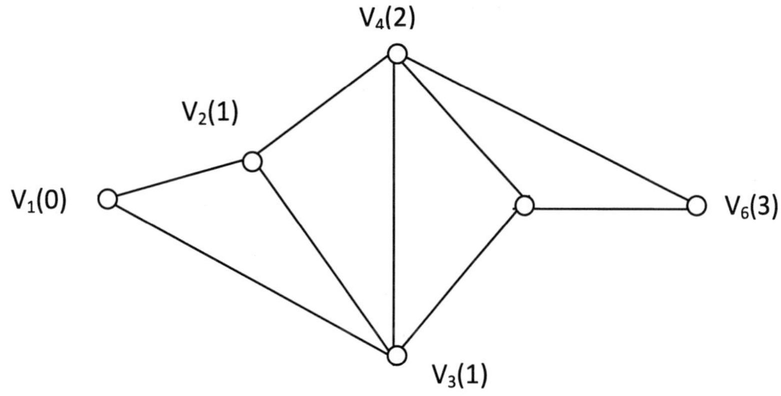

Figure 3.17

Solution

Using BFS, label v_1 as v_0 for i = 0, then adjacent vertices v_2 and v_3 are labeled i = i + 1 = 0 + 1 = 1. Vertices adjacent to v_2 and v_3 (i.e., v_4 and v_5) are labeled i = i + 1 = 2. Then vertex v_6 is labeled as

$i = i + 1 = 2 + 1 = 3$. Since last vertex v_6 is labeled 3, therefore, the length of a shortest path from v_1 to v_6 is 3.

Now using **BTS** Algorithm, since $\lambda(v_6) = 3$, we start with $I = 3$ and $v_i = v_6$. Vertex adjacent to v_6 are $v_4$and v_5. So we choose $v_4$or v_5 with $\lambda(v_5) = 2$ and assign $v_5 = 2$. Next we choose v_3 adjacent v_5 with $(v_3) = 1$ and assign $v_3 = 1$. Finally, we take v_1 adjacent to v_3 with $\lambda(v_1) = 0$ and assign $v_1 = 0$. Hence, the shortest path from v_1 to v_6 is $v_1v_3v_5v_6$.

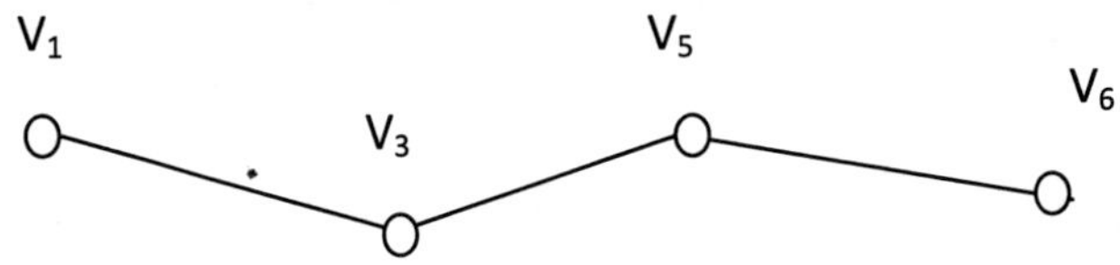

Figure 3.18

3.9.2 Shortest path in a Weighted Graph

Many problems can be modeled using graphs with weights (a number denoting distance, cost, flight time, flightfare etc). assigned to their edges. In these problems determining a path of least length between two vertices is a problem. Given a weighted graph G (V, E), we interpret weight as the length of the direct route from vertex v_i to v_j. For any vertices s, t Є V, suppose that v_1, v_2, ...v_n V, and that the edges (s, v_1), (v_1, v_2), ... (v_n, t) provide a path from s to t. The length of this path is the sum of the weights of the edges on this path and the shortest path between s and t is the minimum length of the path. We denote the shortest path from s to t as d(s, t). To be more specific, the length of a path (different from the useof length to denote the number of edges in a path in a graph without weights) in a weighted graph is the sum of weights of the edges of the path. There are several different algorithms that find a shortest path between two vertices in a weighted graph. Methods such as Kruskal's and Prim's algorithm find the most efficient way to traverse an entire graph. However, if the distance(cost) between two given vertices needed to be calculated, an alternative method would be required. The procedure to find shortest path between the first and every other vertex in the weighted graph was developed by **Edsger Wy be Dijkstra**, a Dutch Mathematician. **Dijkastra's algorithm** determines the distances (costs) between a given vertex and all other vertices in the graph. **Dijkastra's algorithmis** almost identical to Prim's. The algorithm begins at a specific vertex and extends outward within the graph, until all vertices have been reached. The only distinction is that Prim's algorithm stores a minimum cost edge whereas Dijkstra's algorithm stores the total cost from a source vertex to the current vertex. More simply, **Dijkastra's algorithm** stores a summation of minimum cost edges whereas Prim's algorithm stores at most one minimum cost edge. **Dijkastra's algorithm** creates labels associated with vertices. These labels represent the distance(cost) from the source to that particular vertex. Labels of Dijkastra's algorithm, are of two types: temporary and permanent. The temporary labels are given to vertices that have not been reached. The value given to the temporary label can vary. A temporary label can be replaced with another label if the shortest distance to the same vertex can be found. At a point when it becomes evident that no better route can be found, the status of the temporary label is changed to permanent.

The algorithm begins by initializing any vertex in the graph (vertex S, for example) a permanent label with the value 0, and all other vertices a temporary label with the value of . The algorithm then proceeds to select the least cost edge connecting a vertex with a permanent label (currently vertex S) to a vertex with a temporary label (vertex B, for example). Vertex B's labelis then updatedfrom a temporary to a permanent label. Vertex B's value is then determined by the addition of cost of the edge with VerticesS's value.

The next step is to find the next least cost edge extendingto a vertex with a temporary label from either vertex S or vertex B (vertex C, for example), change vertex C's label to permanent, and determine its distance to vertex S.

This process is repeated until the labels of all vertices in the graph are permanent.

The steps of the algorithm are summarized below:

Step 1: Set λ (S) = 0, (assign the vertexs a permanent label 0). Assign every other vertex v, a temporary label (where, is a symbol that, by definition, is deemed to be larger than any real number). i.e., set λ (v) = ∞

Also set T = V,

where T = Set of "Unlabelled" vertices

V = Vertex set of G.

Step 2: Let u be a vertex in T for which is minimum.

Step 3: If u = t, stop.

Step 4: For every edge e = uv, incident with u, if v T, i.e., v is from unlabelled set of vertices, change (min λ (v) + ω (e)).

Step 5: Change T to T - [u] (Label u) and go to step 2 to find an "unlabelled" Vertex with minimum label.

Example 8: Apply Dijkastra's algorithm to the graph given below and find the shortest path from A to f.

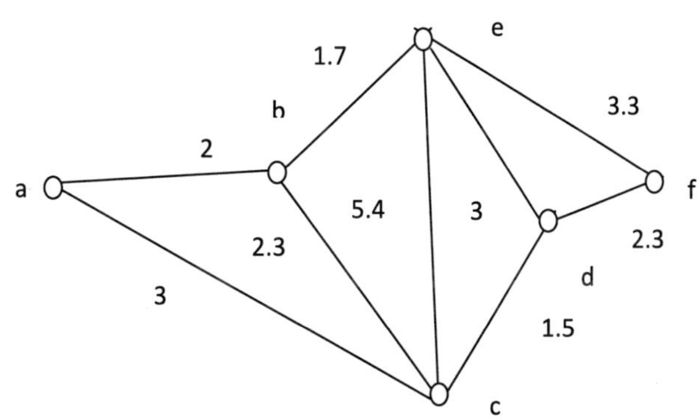

Figure 3.19

Solution

The initial labeling is given by,

Vertex V ab cd ef

Vertex V	a	b	c	d	e	f
λ(v)	0	∞	∞	∞	∞	∞
T	{a	b	c	d	e	f}

Iteration 1: u = a has λ (u) = 0, (minimum). These are two edges incident with a i.e., ab, ac. Both b, c are in T (Unlabelled), λ (b) 2. 0 = 0 + 2. 0 = (a) + ω (ab). So λ (b) becomes 2. 0, Similarly (c) becomes 3. 0. Now, T becomes T-{a} i.e., We label a, thus

Vertex V	a	b	c	d	e	f
λ(v)	0	2.0	3.0	∞	∞	∞
T	{	b	c	d	e	f }

Iteration 2: Again, u = b has λ (u) a minimum and u T. There are 2 edges (bv) incident on b where u T . i.e., be and bc. Since λ (e) = 3. 7 = 2. 0 +1. 7 = (b) + ω (be). So λ (e) becomes 3. 7 and λ (c) = 3. 0 < 4. 3 = 2. 0 + 2. 3 = λ (b) + ω (bc). So λ (c) remains 3. T becomes T-{b} i.e., we label b.

Vertex V	a	b	c	d	e	f
λ(v)	0	2.0	3.0	∞	3.7	∞
T	{		c,	d,	e,	f}

Iteration 3: Again, u = c, has λ (u) a minimum for u in T. There are two edges incident on C from T . i.e., ce and cd. Since λ (e) = 3. 7 < 8. 4 = 3. 0 + 5. 4 = λ (c) + ω (ce). and λ (d) = ∞ > 4. 50 = 3. 0 + 1. 5 = λ (c) + ω (cd). And So λ (e) remains as it is and λ (d) becomes 4. 5 T becomes T-{c} . i.e., we label C.

Vertex V	a	b	c	d	e	f
λ(v)	0	2.0	3.0	4.5	3.7	∞
T	{			d,	e,	f}

Iteration 4: Again, u = e, has λ (u) a minimum for u in T. There are two edges incident on e from T . i.e., ed and ef. Since λ (d) = 4. 57 < 6. 7 = 3. 7 + 3. 0 = λ (e) + ω (ed) and λ (f) = ∞ >7 = 3. 70 + 3. 3 = λ (e) +ω (ef). So λ (d) remains as it is and λ (f) becomes 7. 0 T becomes T- {e}. i.e., we label e.

Vertex V	a	b	c	d	e	f
λ(v)	0	2.0	3.0	4.5	3.7	∞
T	{			d,		f }

Iteration 5: u = d, has λ(u) a minimum for u in T. There is one edge incident on d from T . i.e., df. Since λ (f) = 6. 7 = 4. 5 + 2. 2 = (d) + ω (df). so λ (f) becomes 6. 7. T becomes T- {d} . i.e., we label d.

Vertex V	a	b	c	d	e	f
λ(v)	0	2.0	3.0	4.5	3.7	∞
T	{					f}.

U = f, the only choice. Hence we stop. When the algorithm stops the value λ (v) give the lengths of the shortest paths from the vertex a to each vertexv. Thus, the length of shortest paths from **a to b, c, d, e, f are 2. 0, 3. 0, 4. 5, 3. 7 and 7** respectively.

The following two theorems prove the validity of Dijkastra's algorithm (we are omitting the proof).

THEOREM 3.10

In **Dijkastra's algorithm**, if at some stage is finite (i.e., not) For the vertex vb then there is a path from s to v whose length is λ(v).

THEOREM 3.11

In Dijkastra's algorithm when a vertex u is chosen in step 2, its label λ (u) has value (u). (In particular, upon termination of the algorithm, λ (t) = (t)).

SUMMARY

1. A **walk** is an alternating sequence of vertices and edges, beginning and ending with a vertex in which each vertex is incident to the two edges that precede and follow it in the sequence, and the vertices that precede and follow an edge are the **end vertices** of that edge. A walk is **closed if** the first and last vertices are the same and **open** if they are different. The ***length*** of a walk is the sum of the lengths of its edges. The **path** is any route taken along theedges of the graph. That is, A **path** is an open walk . If the paths originate and terminateat the same point thenit is known asa **circuit.** A graph is **connected** if it is possible toestablish a path from any vertex to any other vertex of a graph. A digraph D is connected(or weakly connected) if the underlying graph D is connected. A digraph is strong if for every path u, v of vertices, D contains a u-v path and a v-u path.

2. A **trail** is a walk in which all the edges are distinct. A closed trail has been called a tour. A trail is **Eulerian** if it uses all edges exactly once. That is, it is a circuit that crosses every edge once.

3. A circuit that visits every vertex exactly once is known as a **Hamiltonian circuit**. In a **Hamiltonian path** each vertex is crossed exactly once

4. **Fleury's Algorithm** for constructing Euler circuitstarts with an arbitrary vertex v of A connected multigraphs and forms a circuit by choosing edges successively. Once an edge is chosen it is removed. Edges are chosen successively so that each edge begins where the last edge ends, so that this edge is not a cut edge unless there is no more alternative.

5. **Hierholzer's algorithm** for constructing Euler circuit, starts with any closed trail in G and "attach" to it, systematically, "detour" trail until all edges of G are used up.

6. **Dijkastra's Algorithm** for a weighted graph proceeds by finding the length of the shortest path from a point in the Graph to destination, spanning all vertices thatare reachable from s (i.e., from source to every other vertex in the graph).

EXCERCISES

1. In the graph shown below, identify (1) walk of length 7 that is neither a trail nor a path (2) A closed walk that is not a circuit (3) A circuit that is not a cycle (4) closed walk thatis not a cycle & (5) a3 cycle?

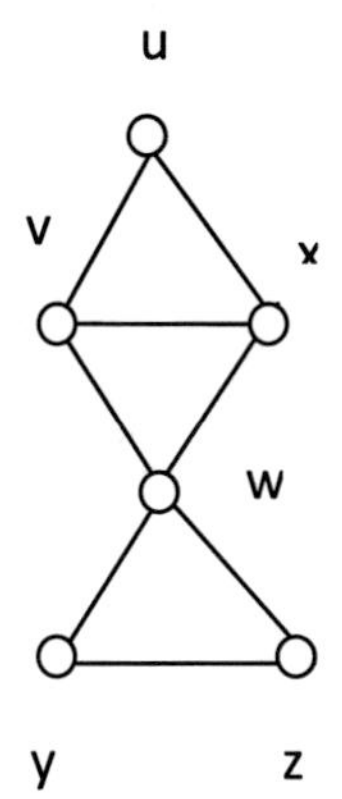

2. Is the graph shown below Eulerian ? Hamiltonian ?

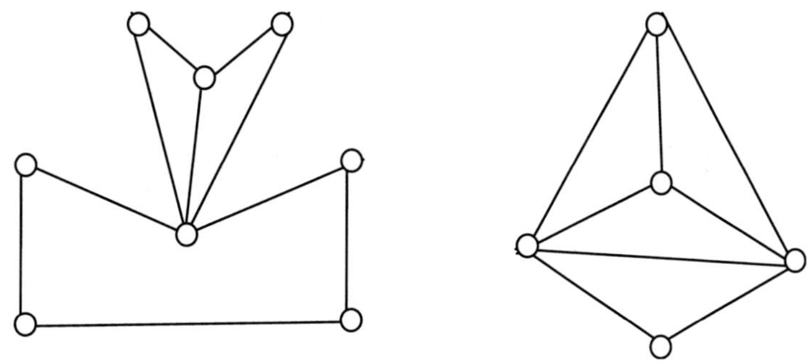

3. Draw a graph with six verticeswhich is Hamiltonian but not Eulerian? And (b) Eulerian but not Hamiltonian?

4. For the graph G find all simple paths from v_1 to v_3 and also find the length of the smallest path? Find all cycles in the graph

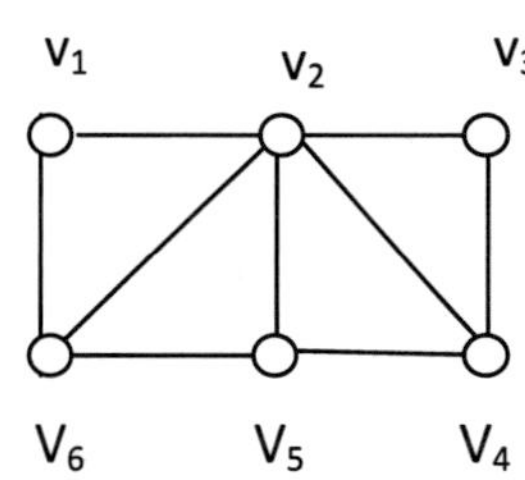

5. Identify the Eulerian trail in the graph shown below

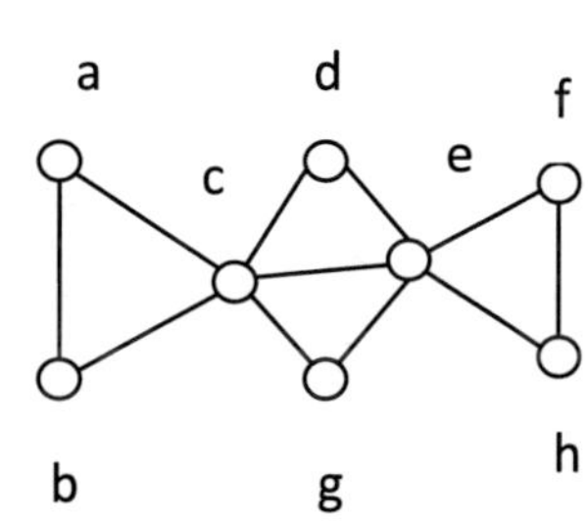

6. Does each of the following list of vertices form a path in the following graph? If so which paths are simple? Which are circuits ? What are the length of the paths?

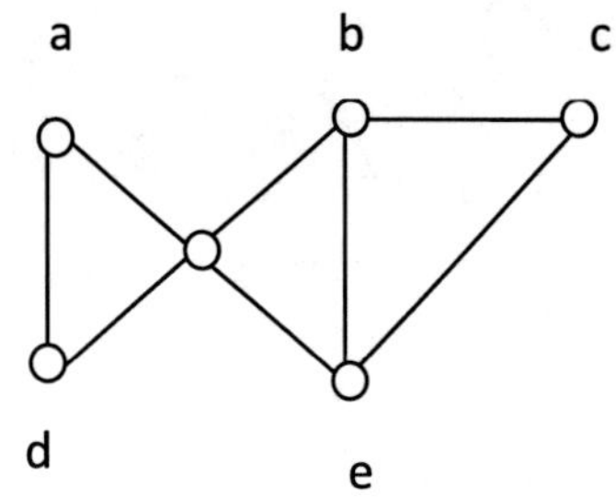

7. Determine whether the given graph has a Hamiltonian circuit ? If it does, find such a circuit? If it does not, give an argument to show why no such circuit exists?

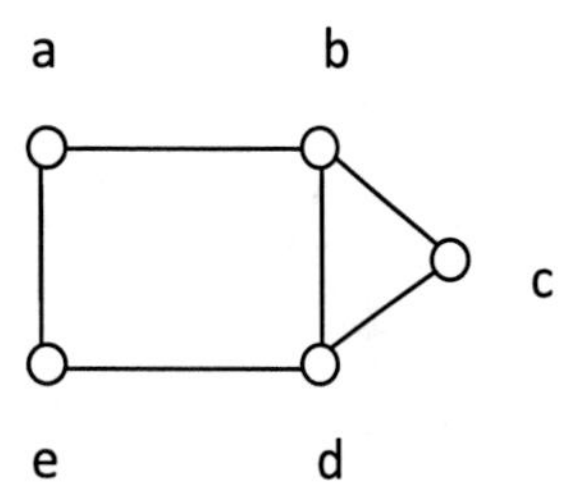

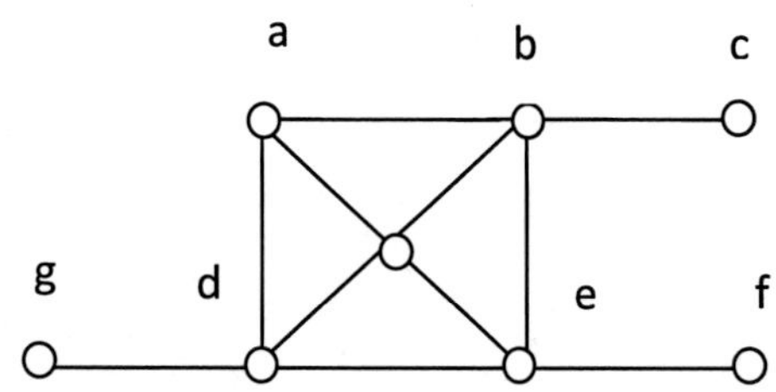

8. Determine whether the directed graphs in the following figures has an Euler path. Construct anEuler path if it exists?

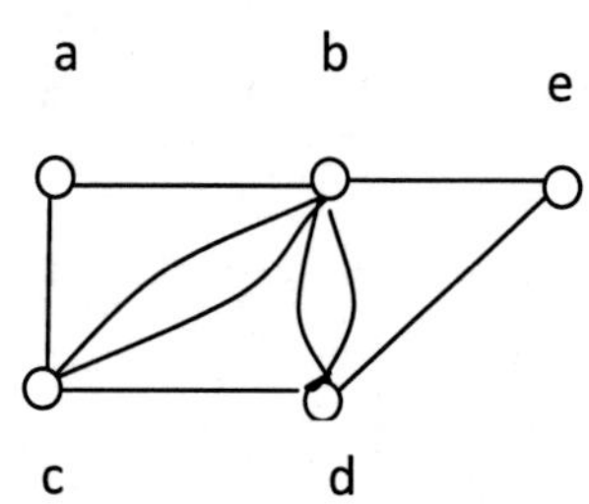

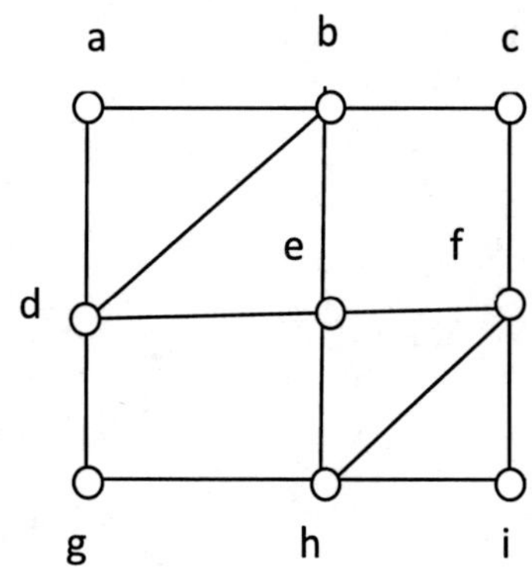

9. Determine whether each graph has an Euler circuit. Construct such a circuit when one exists?

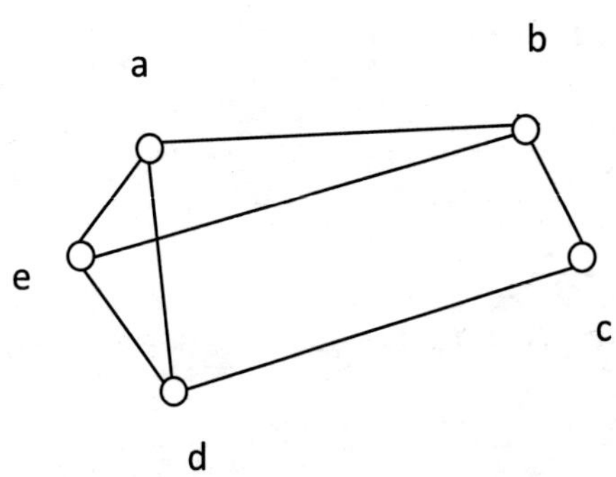

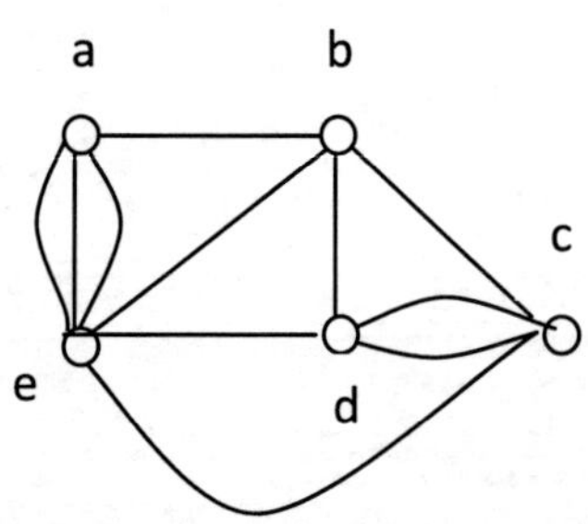

10. Determine whether each directed graph has an Euler circuit. Construct such a circuit when one exists?

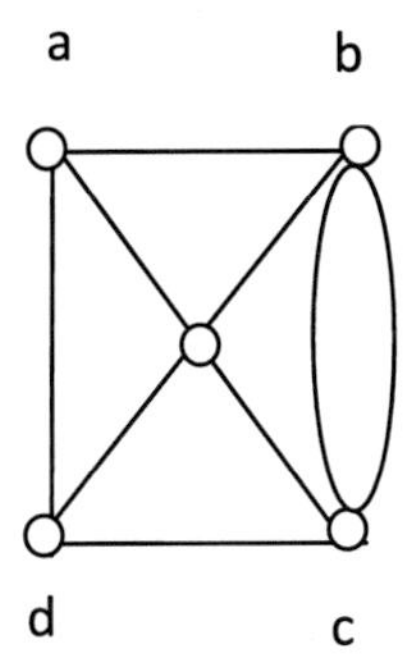

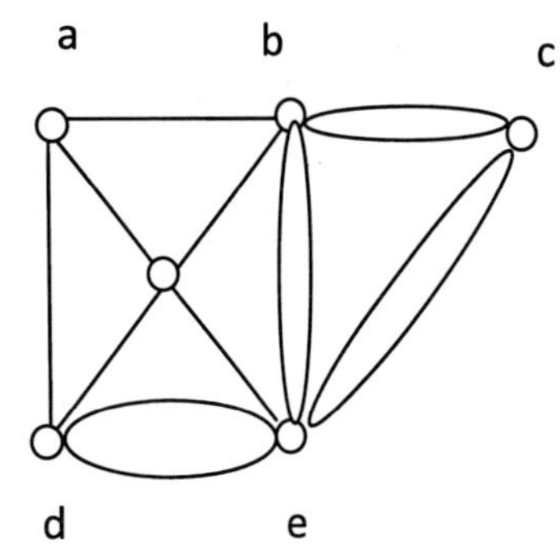

11. For the graph shown in fig determine whether or not each of the following sequences of edges form a path?

    (a) $\{(v_1, v_3), (v_3, v_2), (v_6, v_4), (v_4, v_3),$

    (b) $\{(v_1, v_3), (v_3, v_4), (v_4, v_5), (v_5, v_1)\}$, &

    (c) $\{(v_3, , v_2), (v_2, v_4), (v_4, v_6)\}$? Find all cycles in the graph?

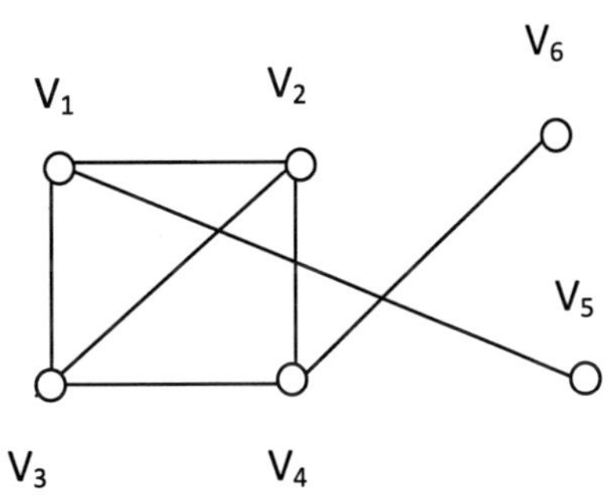

12. For the graph G find all simple paths from v_1 to v_3 and also find the length of the smallest path? find all cycles in the graph?

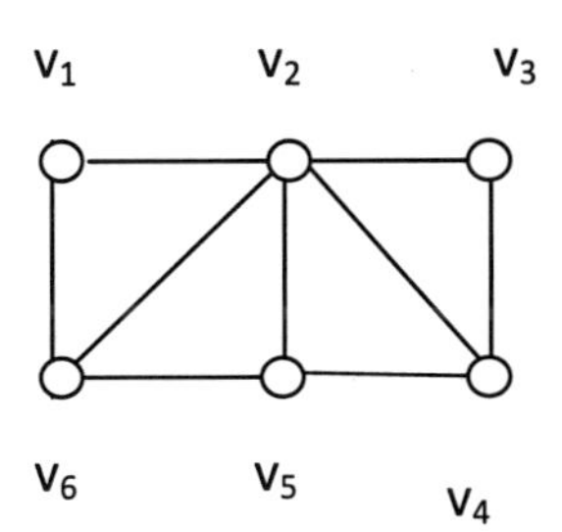

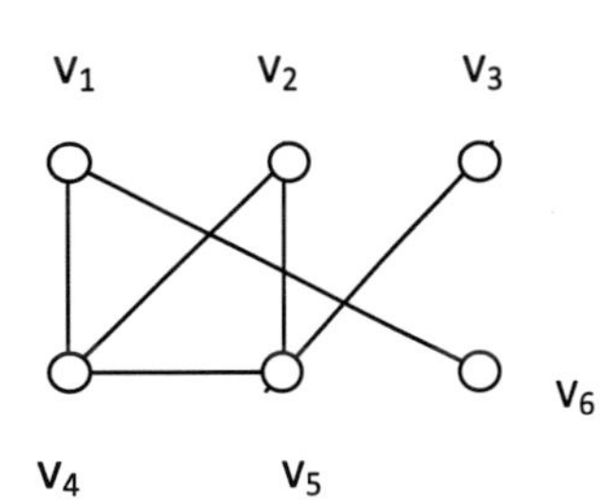

13. Use Fluery's algorithm on the graph below to find an Euler circuit

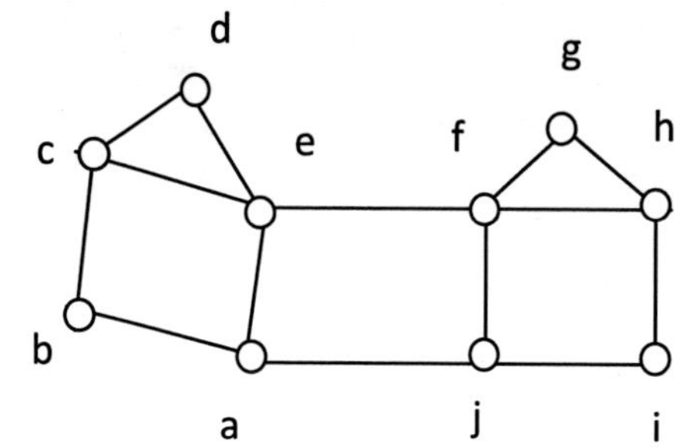

14. Apply Dijkastra's Algorithm to find the shortest path from s to t in the graph given below?

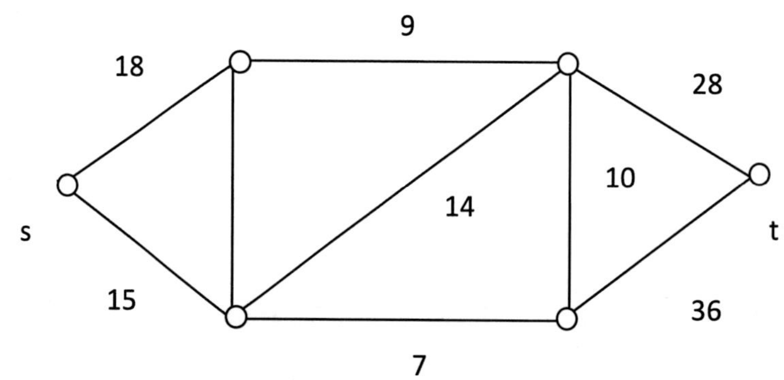

15. Apply Dijkastra's Algorithm to find the shortest path from a to f in the graph given below:

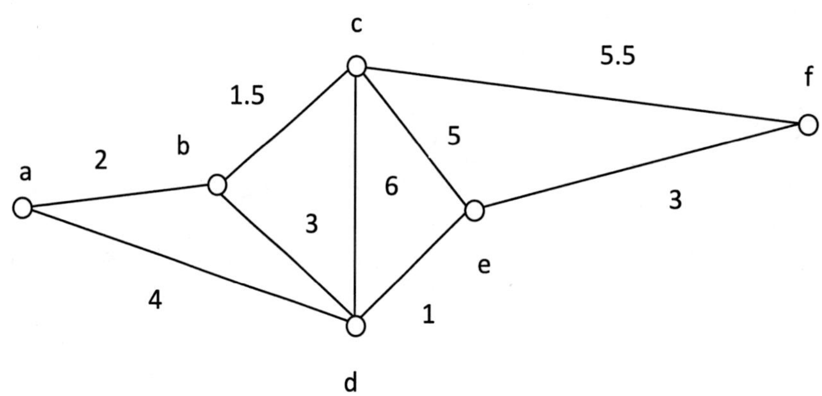

16. Carry out two optimal method and the closest insertion method for the travelling salesman problem for the complete weighted graph ?

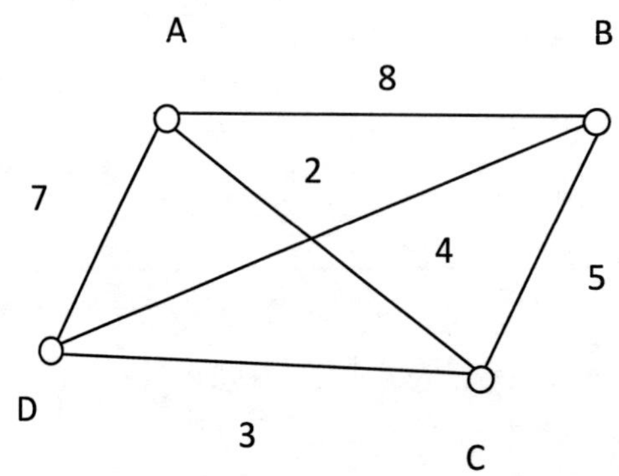

4

TREES

4.0 INTRODUCTION

Another very simple and important graph is a Tree. There are several forms of trees and can be applied for solving a wide variety of problems in Computer science applications such as organizing and sorting of data in a database, Coding theory, language compilers, etc. Trees can either be undirected or directed graphs. Trees are important not only for sake of their applications to many different fields, but also to graph theory itself. The very simplicity of trees makes it possible to investigate conjectures for graphs in general by studying the situation for trees.

In the first half of this chapter we will discuss various types of undirected trees, that too, without certain paths called cycles and their properties. Obviously, then all Multigraphs cannot be considered as trees. Other graph theoretic terms related to trees will also be discussed.

The connectivity of graphs is particularly intuitive area of graph theory and extends many concepts. Using geometric terminology we study connectivity of trees in the second part. This is followed by a discussion of spanning tree of a graph G, and mention the dual construction of a collection of independent co cycles from each spanning tree. . Finally, two graph theoretic algorithms for constructing an optimal spanning tree are also discussed.

4.1 TREES

A graph is called acyclic if it contains no cycles. That is, there is one and only one node (vertex) to any other node. Since loops are cycles of length one while a pair of parallel edges produces a cycle of length two, any acyclic graph must be simple. A graph G is called a tree if it is a connected acyclic graph. When dealing with trees, we often use T rather than G to denote a tree. The graphs $G_{1,}$ G_2 & G_3 shown in Figure 4.1, (a) for instance is a tree, while G_4 & G_5 shown in Figure 4.1

(b) is not a tree, because it contains the cycles. Hence a graph with a unique path between any two vertices is a tree. Trees with one, two, three, and four vertices are shown in Figure 4.2.

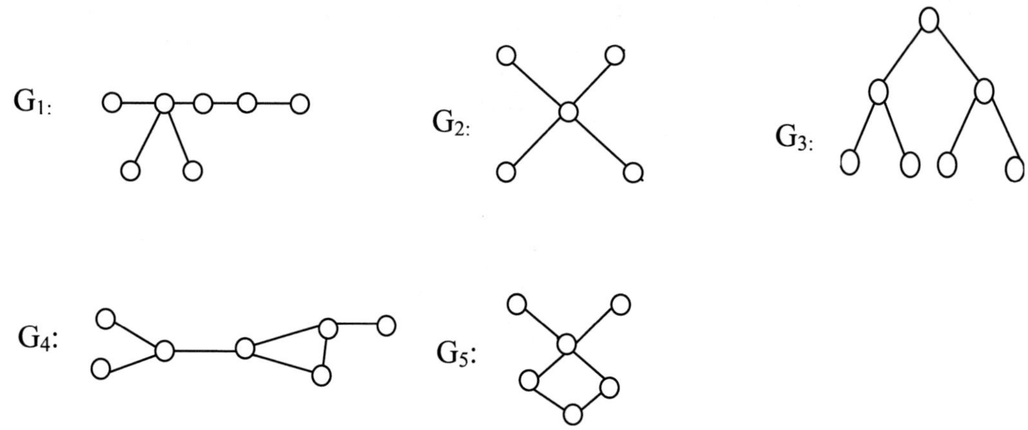

Figure 4.1 **(a)** Graphs G_1, G_2& G_3 –Trees **Figure 4.1 (b)** G_4 & G_5-Not a Tree.

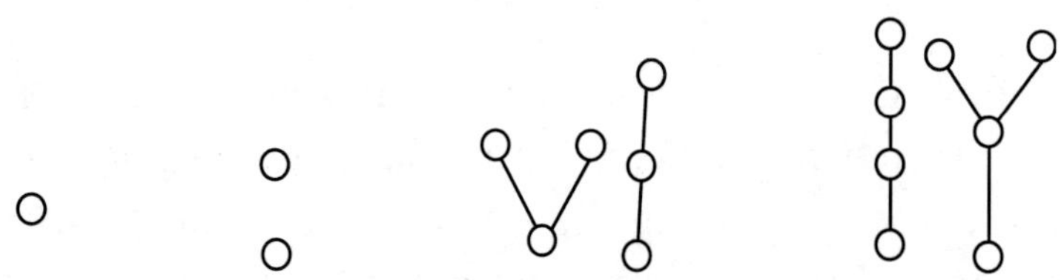

1 vertex 2 vertex 3 vertex 4 vertex

Figure 4.2

Figure 4.3 shows all six trees of order 6. The tree $T_1 \cong K_{1,5}$ is a star and $T_6 \cong P_6$ is a path. The number of end vertices in the trees of Figure 4.3 ranges from 2 to 5. A tree containing exactly two vertices that are not end vertices is called a double star. The trees T_2 and T_3 in Figure 4.3 are double stars. A Tree T with one vertex is called a trivial tree; Otherwise Tree T is a nontrivial tree. .

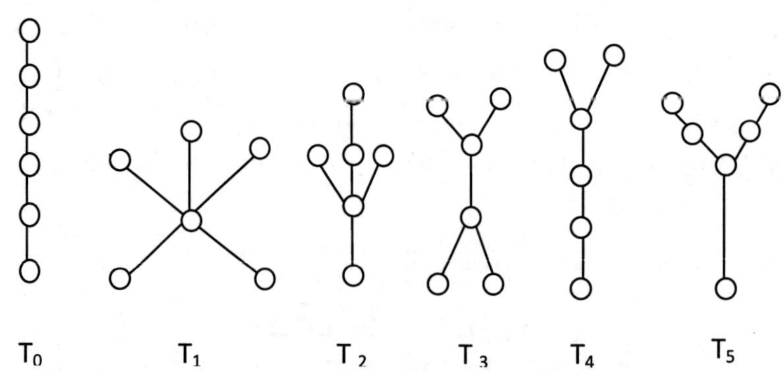

Figure 4.3 The trees of order 6

Note: Trees are often also known as open graphs.

4.1.1 ROOTED TREES

A rooted tree is a tree with a designated vertex v_0 called as root and satisfies following properties:

(i) There is no cycles in the tree.

(ii) The root of the tree is unique.

(iii)Each vertex in the tree other than its root, has in deg(v_1) = 1 and root has in deg(v_0) = 0.

Figure 4.4 is an example of a rooted tree where vertex v_0 is the root of the tree that appears on the top of the tree(. ie. , tree grows downwards.).

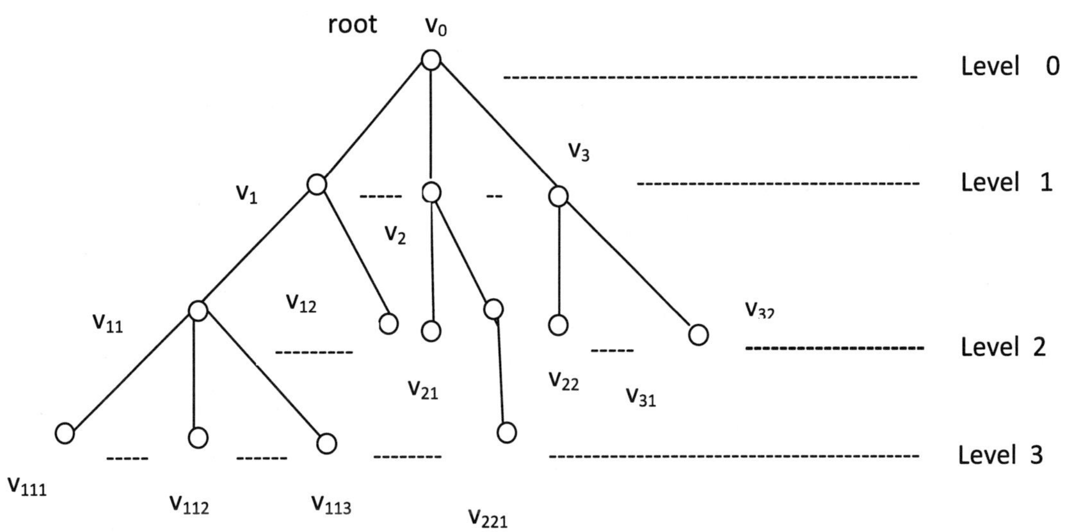

Figure 4.4 Rooted Tree

The edges in a rooted tree define Predecessor-Successor (parent –child) relationship. The *level* of a vertex *v* in a rooted tree is the length of the simplest path from the root. In Figure 4.4, vertex v_0 is the predecessor of vertices v_1, v_2, and v_3. These vertices are called *level* 1 vertices, while v_0 is said to be at level o and v_0 is sometimes called the *parent* of these level 1 vertices. Vertices $v_{11}$and v_{12} are called *offspring* (children) of vertex v_1, while vertices v_{21} and v_{22} are called children of vertex v_2. The vertices having the same parent are called also called *sibling*s of each other. Each vertex at level 1 has no other edges entering it, but vertices v_1, v_2, and v_3 have edge or edges leaving it. The edges leaving a vertex at level 1 are drawn downward and terminate at various vertices such as $v_{11,}$ v_{12}, v_{21}, v_{22}, v_{31} and v_{32}. which are said to be at level 2, vertex v_2 is called the parent of the two offspring v_{21} and, v_{22}. This process continue for as many levels as required to complete the tree.

The descendent of a vertex v_i consists of the vertex itself, all its children together with descendents of those children. In Figure 4. 4, for example, vertices v_{221}, v_{21}, v_{22} are considered descendent of v_2 and v_0. In a tree all vertices are descendents of the root vertex.

The vertices of the tree that have no offspring are called the *leaves* of the tree. (i.e., vertices of degree 1). A directed path from any vertex to a leaf is called a *branch.* Vertices v_{111}, v_{113}, v_{112}, and v_{221} etc. are *leaves* in Figure 4. 4. The path (v_0, v_1, v_{11}, v_{112}), (v_0, v_2, v_{22}, v_{221}), etc are *branches.* The *path length* of a vertex in a rooted tree is the number of edges in the path from root to the vertex. That is, the vertices which has no children, i.e., out deg(v_i) = 0, are called *external* (*leaf or terminal*) vertices, otherwise *branch* (non terminal) nodes.

The height of a tree is the maximum level of any leaf in the tree. This equals the length of the longest path from root to any leaf. The tree in Figure 4.4 is of height 3.

A vertex n_1 is called an *ancestor* of vertex n_2 (n_2 is a *descendent* of n_1), If n_1 is either the parent of n_2 or the parent of some ancestor of n_2. For example in Figure 4.4, v_0 is the ancestor of v_{221} & v_{221} is the descendent of v_2, but v_{11} is neither an ancestor nor a descendent of v_2. The definition of tree mentioned above applies to undirected graphs as well as to directed graphs.

A rooted tree is a *directed* tree if there is a root from which there is a directed path to each vertex of the tree. In a directed tree, a particular vertex is called a root if this vertex is of degree 0. A tree together with its root produces a graph called a rooted tree. In a directed root, there is exactly one root. However, in a tree any vertex may be designated as root. Note that in the rooted tree, every edge is directed away from the root.

Example: 1

Consider the rooted tree shown in figure 4.5. Identify (a) the path length fromits root to each of the vertices v_1, v_6, v_8 & v_{13}? (b) all vertices. which precede v_{11}. (c) All vertices which follow v_1. (d) The children of v_2(e) The sibling of v_3 ?.

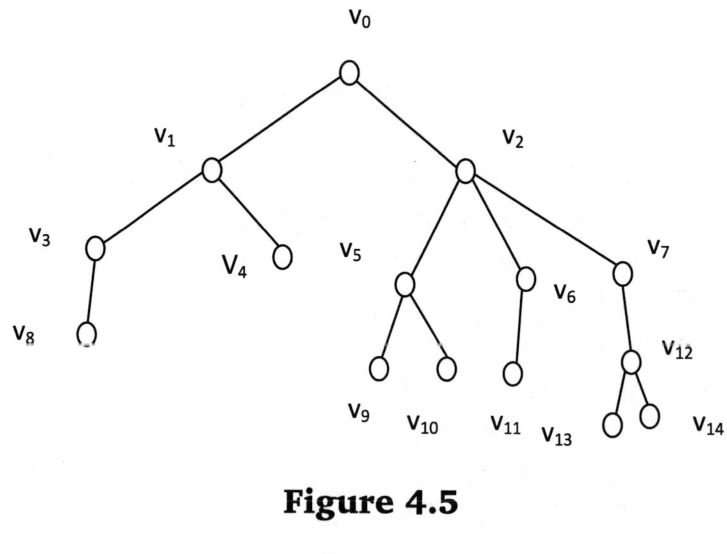

Figure 4.5

Solution

The path length in the rooted tree as shown in Figure 4. 5 from its root to each of the vertices v_1, v_6, v_8 & v_{13} is obtained by listing the root between these two vertices including the vertex itself as follows :

(a) From v_o to v_1: (v_o, v_1) path length is 1.

From v_0 to v_6: (v_0, v_2, v_6) path length is 2.

Fromv$_0$ to v_8: (v_0, v_1, v_3, v_8) path length is 3.

Fromv$_0$ to v_{13} $(v_0, v_2, v_7, v_{12}, v_{13})$ path length is 4.

(b) The path from root v_0 to v_{11} includes the vertices v_0, v_2, v_6 & v_{11}. Hence v_0, v_2, v_6, precede v_{11}.

(c) All paths from vertices v_1 include the vertices v_3, v_4, v_8. That is, v_3, v_4, v_8 follow v_1.

(d) The vertices v_5, v_6 & v_7 are the children of vertex v_2.

(e) The parent of v_3 is v_1 and the children of v_1 & v_3 and hence v_4 is the only sibling of v_3.

4.1.2 BINARY TREES

A rooted tree T is called an ***m-ary*** tree when each internal vertex has at most m children. In particular, if each internal vertex of a rooted tree has at most two children, then such a tree is called *2-ary binary tree*. (Figure 4.6). A vertex of degree one is called pendent vertex. Every binary tree is also a rooted tree having following properties:

(i) The number of vertices in a binary tree is always odd.

(ii) If the number of vertices, say, p in a binary tree is of degree 1 (pendent vertex), Then n-p-1 is the number of vertices of degree 3. That is, there is at least 2 vertices of degree 1 and the number of external vertices is one more than the number of internal vertices.

(iii)A tree consisting of no node (null tree or empty tree) is also a binary tree.

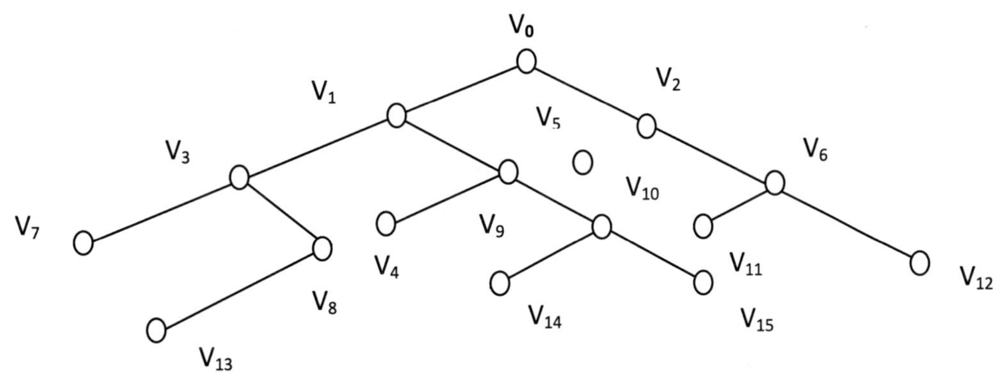

Figure 4.6 Binary tree

An ordered rooted tree is a rooted tree where the children of each internal vertex are ordered. Ordered rooted trees are drawn so that the children of each internal vertex are shown in order from left to right.

In an ordered binary tree (usually called just Binary tree), if an internal vertex has 2 children, the first child is the left child and the second child is called the right child. The tree rooted at the left child of a vertex is called the left sub tree of this vertex, and the tree rooted at the right child of a vertex is called the right sub tree of the vertex.

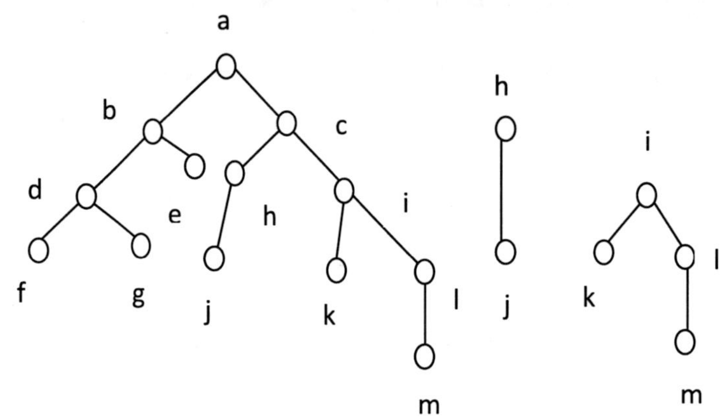

Figure 4.7 A binary tree with left and right sub trees of vertex c.

In the binary tree shown in Figure 4.7, the left child of d is f and the right child is g. The left and right sub trees of c are shown in Figure 4. 7(b) & (c).

Remarks

(i) Number of pendent vertices in a tree is equal to (n+1)/2, where n is the number of vertices in a binary tree.

(ii) In every non-leaf vertex in a binary tree has non empty left and right sub trees, the tree is called as a strictly Binary tree.

(iii) Unlike a general tree that is non –empty, a binary tree can be empty.

4.1.3 COMPLETE BINARY TREE

If every vertex (node) of a rooted tree has m offspring's (children), then such a tree is said to be a complete m-ary tree. In particular, if m = 2, then such a tree is called a complete binary tree. The height of the complete binary tree having m vertices is $h_m = \log_2 m+1$.

All vertices in a Binary tree having two children are called internal vertices and the vertices having no children are called external vertices or leaves. The total number of vertices in a complete binary tree of height h equals the sum of the number of vertices at each level. The significance of complete binary tree is that it is the binary tree with the maximum number of vertices at a given height. Although a complete binary tree contains many vertices, the distance from the root to any leaf (tree's height) is relatively small.

In other words, a rooted binary tree of height, h is called a complete binary tree provided

(i) Any vertex at a level less than h-1 has two children.

(ii) For any vertex v_h in the tree with a right descendent at level h, v_h must have a left child and every left descendent of v_h is either a leaf at level h or has two children.

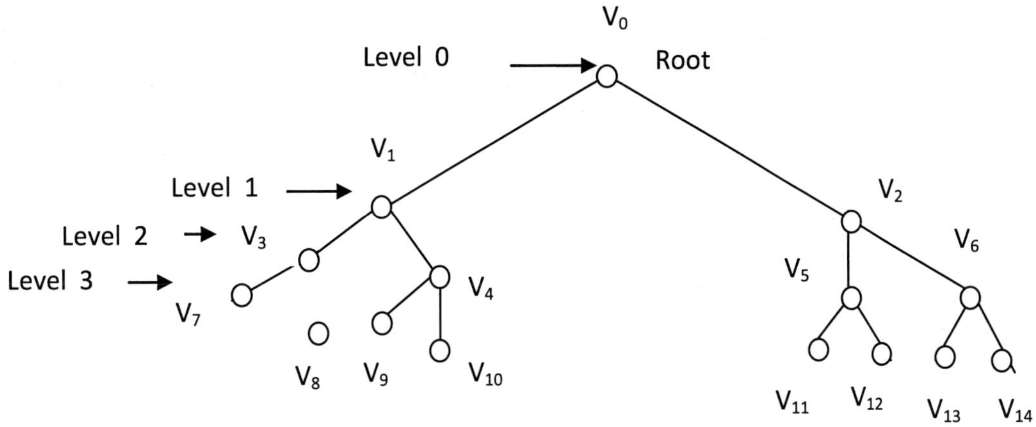

Figure 4.8 A complete binary tree of height 3

For the a complete binary tree of Figure of 4.8, the number of vertices at level 1, 2 &, 3 are 2, 4, and 8 respectively.

4.1.4 BALANCE TREE

A rooted m-ary tree of height h is said to be balanced if all its leaves are at level h or h-1.

4.1.5 CATERPILLARS

Another common class of trees consists of the "caterpillars". A caterpillar is a tree of order 3 or more, the removal of whose end vertices produces a path called the spine of the caterpillar (Figure 4.9(b)). Thus every path, and star (of order at least three) and every double star is a caterpillar, as is every tree shown in Figure 4. 3. The tree T'&T''of T Figure 4.9 are also caterpillars, but on the other hand is T is not a caterpillar. As the number of vertices increases the drawing to tree become complicated.

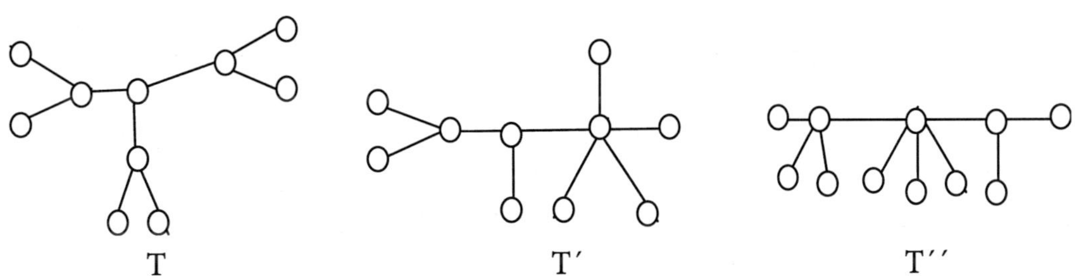

Figure 4.9(a) Tree that is not a caterpillar **Figure 4.9(b)** Two caterpillars

4.1.6 FOREST

Let G be an acyclic graph. Then any sub graph of a graph G must also contains no cycles. The connected components of G are also acyclic and so they are trees. This is why an acyclic graph is also called a forest. A graph with no cycles, where each node has one or no predecessors is called a forest. Any forest consists of a collection of disjoint trees obtained by deleting the root and the corresponding edges connecting the vertices from a given tree (Figure 4.10)

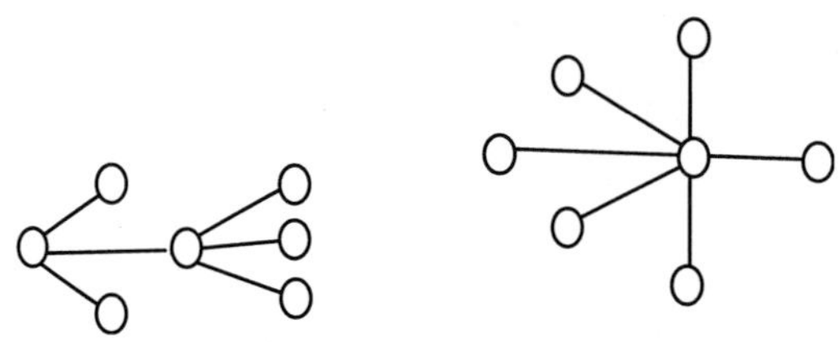

Figure 4.10 Graphs depicting Forests

A tree may be regarded as a forest in which only a single node (root) has no predecessors. Of course, the one fact that distinguishes trees from forests is that a tree is required to be connected, while a forest is not required to be connected.

Trees are often defined as undirected graphs with the property that there is a unique path between every pairs of vertices. Theorem 4. 1 shows that the alternative definition is equivalent to our definition.

THEOREM 4.1

An undirected graph is a tree if and only if there is a unique path between any two of its vertices.

Proof: First assume that, T be a tree. Since T (V, E) is connected, there existat least one path between each pair of distinct vertices in T. Suppose, there exist two distinct paths between a pair of vertices *u* & *v* in T, then some of the edges from thesepaths will form a cycle in T contradicting that T is a tree(Since by definition, a tree is a connected acyclic graph). Therefore, T cannot be a tree.

4.2 SOME IMPORTANT PROPERTIES OF TREES

We have already given one characterization of tress in addition to definition. There are several other characterizations. We list more at this point. As we have already observed, each tree in Figure 4. 5 has two or more end vertices All nontrivial trees have this property.

THEOREM 4.2

Every nontrivial tree has at least two end vertices.

Proof: Let T be a nontrivial tree and among all paths in T, let P be a path of greatest length. Suppose that P is a u-v path, say P : $u = u_0, u_1, \ldots\ldots, u_k = v$, where $k \geq 1$. To prove the theorem, we have to show that u and v are end vertices of G. Necessarily, neither u nor v is adjacent to any vertex not on P, for otherwise, a path of greater length would be produced. Certainly, u is adjacent to u_1 on P and v is adjacent to u_{k-1} on P. Moreover, since T contains no cycles, neither u nor v is adjacent to any other vertices in P. Therefore, deg u = deg v =1.

One major consequence of this result is that if T is a tree of order $k + 1 \geq 2$, then for each end vertex v of T, the sub graph T-v is a tree of order k. This fact is useful for induction proofs on results concerning trees.

We will often need results relating to number of edges and vertices of various types of trees. We illustrate this idea now by showing that the size of every tree is one less than its order, another useful property of trees T.

THEOREM 4.3

If T is a tree with n vertices then it has precisely (n-1) edges.

Proof: We complete the proof the theorem by mathematical induction on the number of vertices n of T . It is easy to see that the theorem is true for n = 1, 2 and 3. Assume that the theorem holds for all trees with fewer than n vertices.

Let T be a tree with n vertices. In T let e_k be an edge with end vertices u_i and u_j. By Theorem 4.1, there is no other path between u_i and u_j except e_k. Therefore, removal of e_k from T will disconnect the graph. Further more, T- e_k consists exactly two components, and since there were no circuit in T to begin with, each of these components is a tree. Both these trees, T_1 and T_2, have fewer than n vertices each, and therefore, by the induction hypothesis, each contains one less edge than the number of vertices in it. Thus T- e_k consists of n-2 edges. Hence T has exactly n-1 edges.

THEOREM 4.4

Every forest of order n with k components has size n-k.

Proof: Suppose that the size of a forest F is m. Let $G_1, G_2, \ldots. G_k$ be the components of F, where $k \geq 1$. Furthermore, suppose that G_i has order n_i and size m_i for $1 \leq i \leq k$. Then n= $\sum_{i=1}^{k} n \sum_{i=1}^{k} n_i$ and m = $\sum_{i=1}^{k} m$ $\sum_{i=1}^{k} m_i$. Since each component G_i $(1 \leq i \leq k)$ is a tree, it follows by Theorem 4. 3 that $m_i = n_i - 1$. Therefore,

$$m = \sum_{i=1}^{k} m \sum_{i=1}^{k} m_I = \sum_{i=1}^{k} n \sum_{i=1}^{k} n_i - 1 = n-k.$$

Next, the following fact about trees prove useful.

THEOREM 4. 5

The size of every connected graph of order n is at least n-1.

Proof: It is easy to see that the Theorem is true for connected graphs of order 1, 2, or 3. Assume that the theorem is false However. Then there exists a connected graph G of smallest order n whose size m is at most n-2. Necessarily, $n \geq 4$. Since G is nontrivial connected graph, G contains no isolated vertices.

We claim that G contains an end vertex; for assume, to the contrary, that the degree of every vertex of G is at least 2. Then the sum of the degrees of the vertices of G is $2m \geq 2n$; so $m \geq n \geq m+2$, which is impossible. So as we claimed, G contains an end vertex.

Let v be an end vertex of G. Since G is connected, has order n_i and size $m \leq n-2$, it follows that G-v is connected and has order n-1 and sizem-1 $\leq$ n-3, contradicting the assumption that G is a connected graph of smallest order whose size is at least 2 less than its order.

THEOREM 4.6

Any tree T with at least two vertices has more than one Vertex of degree 1.

Proof: Suppose T is a tree with n vertices $v_1, v_2, \ldots\ldots, v_n$ and n-1 edges. By theorem $\sum_{i=1}^{n} \deg v_i = 2$ (n-1) = 2n-2. We already know that some vertex v_1, has degree 1. If the remaining n-1 vertices each have degree 2 or more, the sum of all degrees would be at least 1+2(n-1) = 2n-1, a contradiction. So there has to be another vertex, in addition to v_1, that also has degree 1. Hence the theorem is proved.

Corollary 4.7

If $\delta(G) \geq \delta(G) \geq 2$, then G contains a cycle.

Proof: If G has no cycles, then G is a forest and hence $\delta(G) \leq \delta(G) \leq 1$ by the above theorem.

THEOREM 4.8

In every non trivial tree there is at least one vertex of degree 1.

Proof: We Start at any vertex, say, v_1. If $\deg(v_1) \neq 1$, move along any edge to a vertex v_2 incident with v_1. If $\deg(v_2) \neq 1$, continue to a vertex v_3 along a different edge. We can continue this process to produce a path v_1-v_2-v_3-v_4, ... None of this v_i's (nodes) are repeated in this path since then we would have a circuit- which a tree may not have. Since the number of vertices in a graph is finite, this path must end some where. Where it ends must be a vertex of degree 1. Since we can enter this vertex but cannot leave it. That is, every non trivial tree has is at least one vertex of degree 1.

THEOREM 4.9

A graph with n vertices and less than n-1 edges cannot beconnected.

Proof: Let G be a circuit less graph with n vertices and n-1 edges and is disconnected. In that case G will consist of two or more circuit less components. Further, Let G consist of two components G_1 and G_2. Add an edge e between a vertex v_1 in G_1 and $v_2$in G_2. Since there was no path between v_1 and v_2 in G, adding an edgee did not create a circuit. Thus G - Ue is a circuit less, connected graph of n vertices and e edges, which is not possible, because of Theorem 4.1.

THEOREM 4.10

A tree with at least two vertices contains at least two pendant vertices.

Proof: Consider a largest path of a tree T. The end vertices of this path must be pendant vertices of T; otherwise, the path extendable to a larger path or else T contains a cycle, a contradiction.

THEOREM 4.11

A simple undirected graph is a tree if G is connected and contains no cycles.

Proof: Suppose that G is a tree. Since each pair of vertices are connected by a path, G is connected. If G contains a cycle containing distinct vertices u and v, then u and v are joined by at least two simple paths, the one along one portion of the cycle and the other path completing the cycle. This contradicts the assertion that there is a unique simple path between u and v, and thus a tree has no cycles.

Conversely, suppose that G is a connected acyclic graph. Let u and v be any pair of vertices of G. If there are two different simple paths, P_1 and P_2, from u to v, then we can find a cycle in G as follows: Since P_1 and P_2 are different paths there must be a vertex x, on both paths such that the vertex following v_1 on P_1 is not the same as the vertex following v_1 on P_2. Since P_1 and P_2 terminate at v, there is a first vertex after v_1, say v_2, which P_1 and P_2 have in common. Thus, the part of P_1 from v_1 to v_2 together with that part of P_2 from v_1 to v_2 form a cycle in G. This contradicts the assumption that G has no cycles. Therefore, G has exactly one path joining u and v.

THEOREM 4.12

If two non adjacent vertices of a tree T are connected by adding an edge, then the resulting graph will contains a cycle.

Proof: If T has n vertices T has n-1 edges and then if an additional edge is added to the edges of T the resulting graph G has n vertices and n edges. Hence G cannot be a tree by Theorem 4.1. However, the addition of an edge has not affected the connectivity. Hence G must have a cycle.

THEOREM 4.13

If u and v are non adjacent vertices of a tree T, then T + uv contains a unique cycle.

Proof: If P is a unique u-v path in T, then P+uv is a cycle in T+Uv. It is unique, as the path P is unique in T.

Our next characterization is trees involving cut edges.

Theorem 4.14

A connected graph G is a tree, if and only if, every edge of G is a cut edge of G.

Proof: If G is a tree, there are no cycles in G. Hence no edge of G can belong to a cycle. By Theorem 4.1, each edge of a connected graph G is a cut edge of G, then G cannot contain a cycle, since the edges of a cycle are not cut edges of G. Hence G is a tree.

Another important feature of a tree is that its vertices are connected together with minimum number of edges. A connected graph is said to be minimally connected if removal of any one edge from it disconnects the graph. A minimally connected graph cannot have a circuit; otherwise, we could remove one of the edges in the circuit and still leave the graph connected. Thus a minimally connected graph is a tree. Conversely, if a connected G is not minimally connected, there must exist an edge e_k in G such that G-e_k is connected. Therefore, $e_k$is in some circuit, which implies that G is not a tree. Hencethe following theorem.

THEOREM 4.15

A graph is a tree if and only if it is minimally connected.

Proof: Let G be a tree. To prove that G is minimally connected, let us assume that G is not minimally connected. That is removal of the edges from the graph does not disconnect it. This implies that the edge is in some cycle, which is a contradiction to the statement that the graph G is a tree.

Conversely, G is minimally connected. Since G is minimally connected, implies G is connected and removal of one edge disconnects the graph. That is G is a tree.

The results of the preceding theorems can be summarized by saying that the following are five different but equivalent definitions of a tree. That is, a graph G with n vertices is called a ***tree*** if

1. G is connected and circuit less, or
2. G is connected and has n-1 edges, or
3. G is circuit less and has n-1 edges, or
4. There is exactly one path between every pair of vertices in G, or
5. G is a minimally connected graph.

4.3 TREE OF AN ALGEBRAIC EXPRESSION

We can represent any algebraic or arithmetic expressions containing operands and binary operators using ordered rooted trees. An ordered rooted tree can be used to represent expressions, where the internal vertices represent operations, the leaves represent variables or numerals. That is, the root of the tree contains an operator that is to be applied to the results of evaluating the expressions represented by left and right sub trees. A vertex representing a n operator is a non leaf, whereas a vertex representing an operand is a leaf.

For example, an algebraic expression E = (a+b)* ((c/d)-e) involving binary operation is represented by the binary tree shown in Figure 4.11

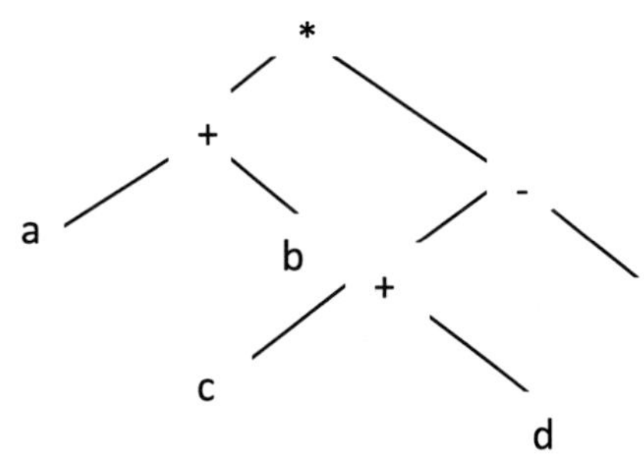

Figure 4.11 Binary tree representation of algebraic expression E= (a+b)* ((c/d)-e).

Example 2:

What is the binary tree that representation the expression ((x+y)↑ 2 + ((x − 4)/3))?

Solution:

First, a sub tree for the expression + is constructed. Then this is incorporated as part of the larger sub tree representing(x+y)) ↑ 2. Also, a sub tree for x-4 is constructed and then this is in corporate into a sub tree representing (x-4)/3. Finally, the sub trees representing (x+y))↑ 2 and (x-4)/3 are combined to form the binary tree representing ((x+y)↑ 2 + ((x − 4)/3)). Now the binary tree is:

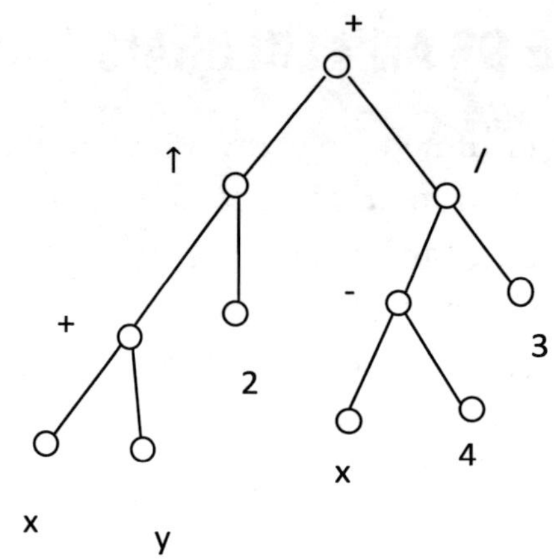

Figure 4.12 Binary tree representation of the algebraic expression:

$$((x+y)\uparrow 2 + \left(\frac{x-4}{3}\right)\uparrow 2 + \left(\frac{x-4}{3}\right)).$$

4.3.1 INFIX, PREFIX AND POSTFIX NOTATION

Infix form: We can represent algebraic or arithmetic expressions in three different ways. They are Infix, Prefix and postfix form of an expression. To get ***infix form*** of an expression we traverse its rooted tree in in order. The original algebraic expression appears with its elements and operations in the order in which the originally occurred. In this form, we include parenthesis in the in order traversal as when an operation appears. For example. The infix notation for x*y for a binary operation is *xy.

Example 3:

Give the infix form of the expression: $(x+xy)+(\frac{x}{y})$. ?.

Solution:

Infix form:((x+(x*y))+(x/y)). The corresponding Binary tree is:

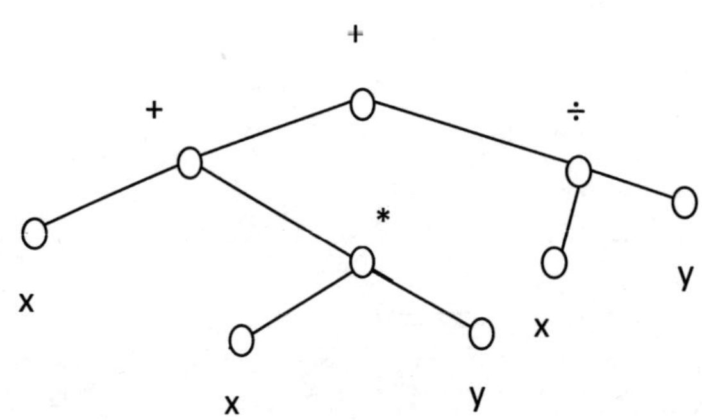

Figure 4.13 The Binary tree for the infix form of expression:((x+(x*y))+(x/y)).

Prefix or polish form: it is an algebraic expression without parenthesis. The procedure to express an algebraic expression in prefix form is as follows: Move from left to right until a string of the form Bxy is found, where *b* represents a binary operator such as, +, -, × , etc, and x and, y are numbers. Evaluate xBy and substitute the answer for the string Bxy. It is called Prefix notation because in this form binary operator symbol precedes the arguments. For example, expression E- $(2x-y)^3$ {(a+(b+c)} can be expressed as : *-*2xy↑ $3 + a \div bc$.

Example 4:

Give the prefix form of the expression: ((x+y))↑ $2 + \left(\frac{x-4}{3}\right)$).

Solution:

prefix form of the expression is: +↑ + **xy2/-x43.** The binary tree representing the expression is:

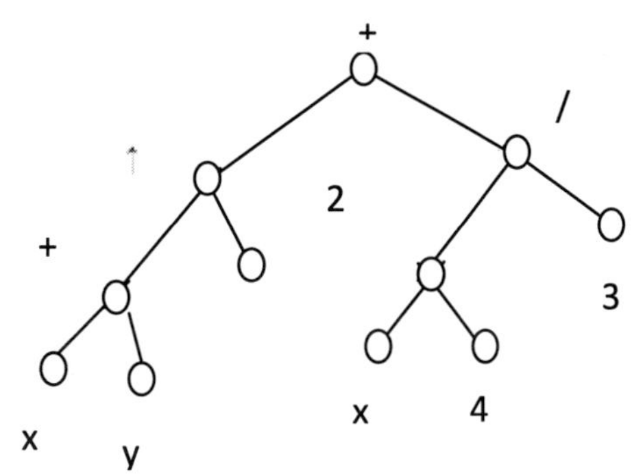

Figure 4. 14 The Binary tree for the Prefix form of expression: (x+y))↑ $2 + \left(\frac{x-4}{3}\right)$).

Postfix or reverse polish form: This form of algebraic expression is same as prefix form except that in this form binary operator comes after its arguments. For example, expression E - $(2x-y)^3$ {(a+(b+c)} can be expressed in postfix form as : 2xy3 ↑ $abc \div + \times$.

Example5

What is the post fix form of the expression: ((x+y))↑ $2 + \left(\frac{x-4}{3}\right)$).

Solution

The post fix form of the expression is : xy+**2 ↑** $x4 - 3$ /+. The corresponding binary tree is:

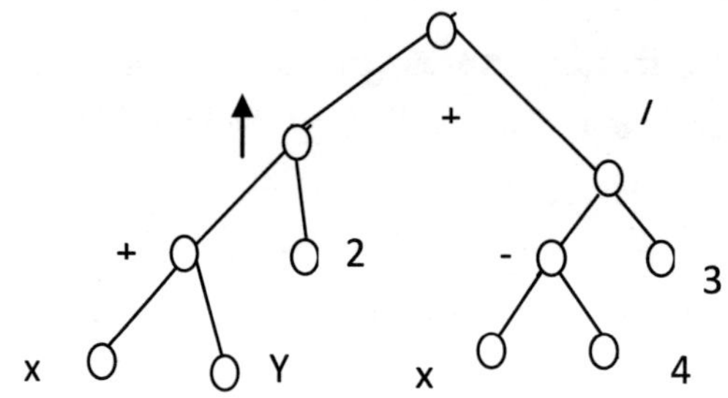

Figure 4.15

The binary tree corresponding to the post fix form of theexpression is : xy+2 ↑ x4 − 3 /+.

4.4 TRAVERSING BINARY TREES

Sometimes, it is desirable to traverse through the tree, enumerating each of its nodes once, in some specified order so as to take some action appropriate to the application, being represented by the tree. The process of visiting and then performing some computation at each vertex of a tree is called searching. A systematic way for visiting every vertex of an ordered rooted tree is called Traversal algorithm. The standard algorithms for searching (traversing) a positional binary tree with root v are:1) Pre order search 2) in order search and 3) post order search

4.4.1 PRE ORDER SEARCH

Let T be an ordered rooted tree with root r. Suppose T has one and only one vertex say r, then r is the pre-order traversal of T. Suppose that T_1, T_2, ……. , T_k are the sub trees at r from left to right in T, then pre-order traversal begins by visiting r. it continues by reversing T_1 in pre order, then T_2 in pre order and so on, until T_k is reached.

The steps of Pre order search algorithm are as follows:

Step 1: First visit the rootnode v of the tree.

Step 2: If V_L (left child) exists, then search the left sub tree T_L of T in preorder.

Step 3: If V_R(right child) exists, then search the right sub tree T_R of T in pre preorder.

4.4.1 IN ORDER SEARCH

Let T be an ordered rooted tree with its roots at vertex r. Suppose T consists only root r, then r is the in order traversal of T. If not, i.e., suppose T has sub trees T_1, T_2, …. . T_k at r from left to right. The in order traversal begins by traversing T_1 in order, then visiting r. It continues by traversing T_2 in order, then T_s in order and finally T_k in order. The steps of this algorithm are as:

Step 1: search the left subtree T_L in order.

Step 2: Visit the root of the tree.

Step 3: search the right sub tree $T_{R'}$ in order.

4.4.2 POST ORDER SEARCH

Let T be an ordered rooted tree with root r. If T has only one vertex r, then r is the post order traversal of T. But if T has sub trees T_1, T_2, …. . T_k at r from left to right, the post order traversal begins by traversing T_1 in post order, then T_2 in post order and so on until T_k and ends by visiting r.

The steps of post order search algorithm are as follows:

Step 1: search left sub tree T_L in post order.

Step 2: search right sub tree $T_{R'}$ in post order.

Step 3: visit the root of the tree. .

4.4.3 CONVERSION OF A TREE TO BINARY TREE

Every tree can be uniquely represented by a binary tree. Now, we show how one can obtain a binary tree which represents a given ordered tree. The steps are as follows:

Step 1: Delete all branches except the leftmost branch in all the vertices.

Step 2: Draw the edges from a vertex to the vertex on the right, of the same parent if any which is at the same level.

Step 3: once this is done, then for any particular vertex, we choose its left and right child in the following manner (i) the left child is immediately below the given vertex and (ii) the right child is on the same horizontal line.

4.5 BINARY SEARCH TREE

A binary tree is called a Binary Search Tree provided vertices in the left sub tree of a vertex v_i are less than the contents of v_i, and all vertices in the right sub tree of v_i are greater than or equal to the contents of v_i. In other words, in BST, every vertex v satisfies following properties:

(i) $v > n$ for every vertex in its left sub tree.

(ii) $v \leq n$ for every vertex n in its right sub tree.

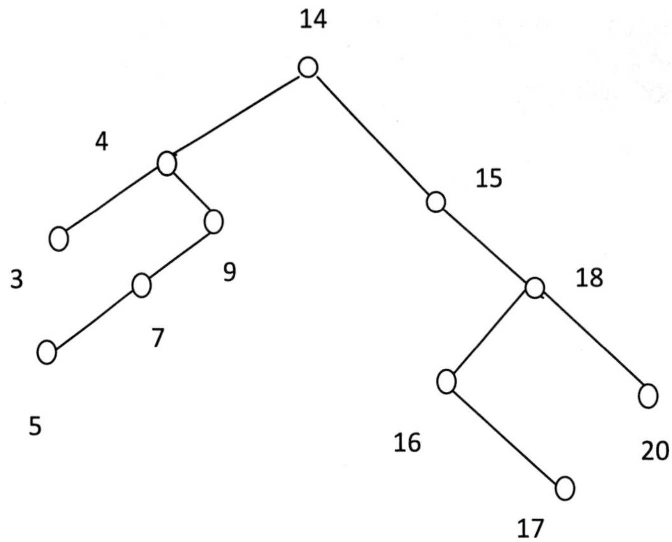

Figure 4.16 Binary Search Tree

It is, however, not necessary that a node in a BST, point to the nodes whose values immediately precede and follow it. Unlike traverse algorithms used to find duplicates, duplicate values are placed in the tree.

4.6 SPANNING TREES

Let G be a graph. A sub graph H of G is called a spanning sub graph of G if the vertex set of H is the same as the vertex set of G. Similarly, a spanning tree of a graph G is a spanning sub graph of G, that is a tree. i.e., a spanning tree of a connected graph G is a sub graph that is a tree and that includes every vertex of G.

The concept of spanning tree exists only for a connected graph G and, if G has n vertices, any spanning tree must necessarily contain n-1 edges. .

Finding a spanning tree in a connected graph G is simple. If G has no cycles, then it is already a tree, so G is a spanning tree for G. If G does contain a cycle, then delete an edge(without deleting any vertex) so as to remove the cycle, but leave the graph connected. By repeating the procedure, we eventually find a connected sub graph without cycles that contains all the vertices of G, that is, a spanning tree. For the connected graph G of Figure, two different spanning trees T_1 & T_2 of G are also shown in Figure.

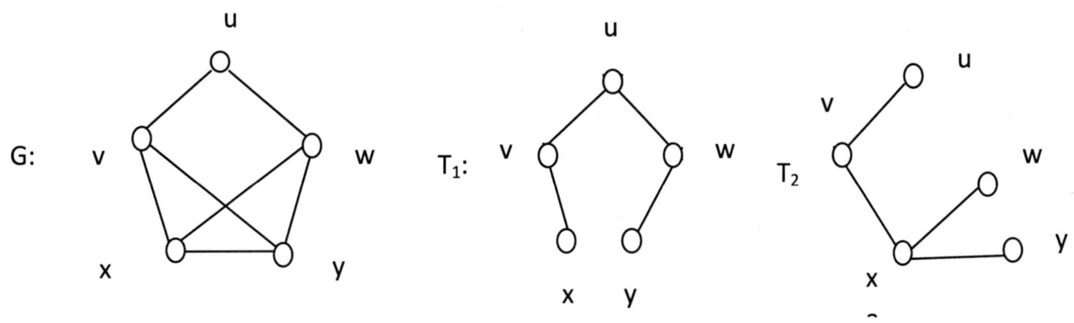

Figure 4.17 Two spanning trees in a graph

THEOREM 4.15

An undirected graph G has a spanning tree if and only if G is connected.

Proof: Suppose that a graph G has a spanning tree T. Let a and b vertices of G. Since a and b are also vertices in T and T is a tree, there is a path P from a to b. However, P also serves as a path from a to b in G; thus G is connected.

Conversely, Suppose that G is connected. If G is acyclic, then G is a tree. Suppose that G contains a cycle. We remove an edge from this cycle. The graph produced is still connected. If it is acyclic, we stop. If it contains a cycle, we remove an edge from this cycle. Continuing in this way, we eventually produce an acyclic, connected sub graph T. which is a tree. Since T contains all the vertices of G, T is a spanning tree of G.

Note: The complete graph K_n has n^{n-2} different spanning trees.

4.6.1 DEPTH – FIRST SEARCHAND BREADTH-FIRST SEARCH

By the previous theorem, an algorithm for finding spanning trees by removing edges from simple cycles, is inefficient. , since it requires that simple cycles to be identified. Instead of constructing spanning trees by removing edges, spanning trees can be built by successively, adding edges. Two algorithms based on this principle will be presented here. They are DF's and BF's . We can build the spanning tree of a connected graph using DF's and BF's. First we describe DF, which is very simple, efficient, and are useful in construction of a spanning tree from a given connected graph. It quickly tests whether a graph is connected, produces a spanning tree in the connected case, and is the key to solving problem about turning graphs into digraphs.

Depth-First Search

We can build a spanning tree of a connected simple graph using DFS. We will form a rooted tree, and the spanning tree will be the underlying undirected graph of this rooted tree. Let G be the given connected graph. Arbitrarily choose a vertex as root. Form a path starting from this chosen vertex by successively adding edges, where each edge is incident with the last vertex in the path and a vertex not already in the path. Continue adding edges to this path as long as possible. If this path consists of vertices of G, this path is the required spanning tree. If not, more edges should be added. Move back to the next to last vertex in this path, and if possible, move back to another vertex in this path (i.e., 2 vertices back from the last) and try again.

Repeat this procedure, beginning at the last vertex visited, moving back up the path one vertex at a time, forming new long paths until no more edges can be added. This process ends with the production of a spanning tree. Each vertex that ends a path at a stage of the algorithm will be a leaf in the rooted tree, and each vertex where a path is constructed stating at this vertex will be an internal vertex.

When this procedure returns to vertices previously visited, it is also called backtracking.

Algorithm DFS

Step 1: (visit a vertex) Let v_1 be the root ofT and set $L = v_1$. (L stands for last visited vertex).

Step 2: (Find an unexamined edge and unvisited vertex adjacent to L). For all vertices adjacent to L, choose the edge (L, v_k), where k is the minimum index such that adding edge (L, v_k), to T does not create a cycle. If no such edge exists, go to step 3; otherwise add edge (L, v_k), to T and set $L = v_k$; repeat step 2 at the new value for L.

Example 6:

Construct a spanning tree for the following graph G

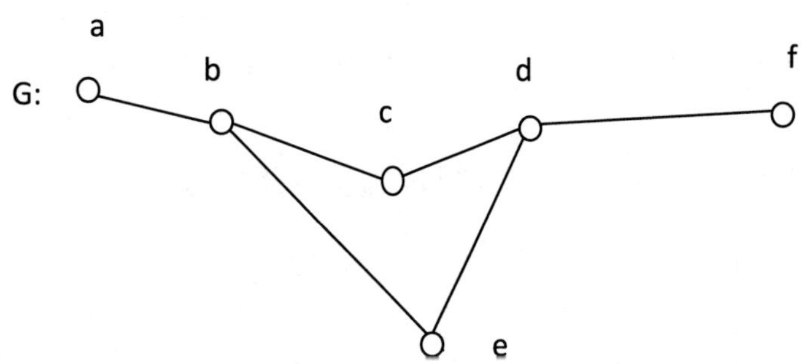

Figure 4.18

Solution

First we choose arbitrarily a vertex say, e as the root. Form a path at e i.e., cdf is the path. Back track to d. Form a path beginning at d in such a way that it has to visit the vertices which were not visited in the previous path. , deba . Since all the vertices of G are visited, this procedure gives thespanning tree T.

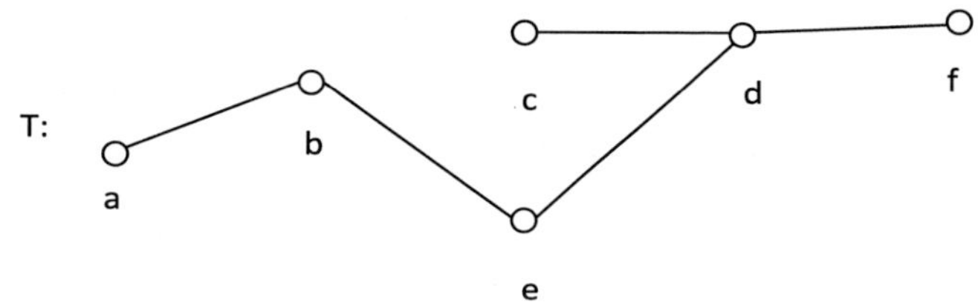

Figure 4.19

Breadth-First Search

We can also produce a spanning tree of a simple graph by the use of BFS. Again, a rooted tree will be constructed, and the underlying undirected graph of this rooted tree forms a spanning tree. First choose a vertex arbitrarily as the root. Add the edges of G which are incident with this vertex. The new vertices added at this stage becomes level 1 in the spanning tree. Order these vertices arbitrarily. Next, for each vertex at level 1 visited in order, add each edge incident to this vertex to the tree as long as it does not create a simple circuit. Order the children of each vertex at level 1 arbitrarily. This produces the vertices at level 2 in the tree. Continue in this manner until all the vertices of G have been added. Ultimately we end the spanning tree.

Algorithm BFS

Step 1: Let v_1 be the root of G. For the set $V = \{ v_1 \}$

Step 2: Add new edges. Consider the vertices of the graph in order, consistent with the original labeling, then for each vertex $x \in V$, add the new edge $\{ x, v_k \}$ to T, where k is the minimum index such that adding the edge $\{ x, v_k \}$ does not produce any cycle. If no edge can be added, then STOP. Then T is a spanning tree for G. After all the vertices of V have been considered in order, go to step 3.

Step 3: replace V by all the children v in T of the vertices x of v where the edges $\{ x, v \}$ were added in step 2. Go back and repeat step 2 for the new set V.

Example 7:

Construct a spanning tree of the following graph

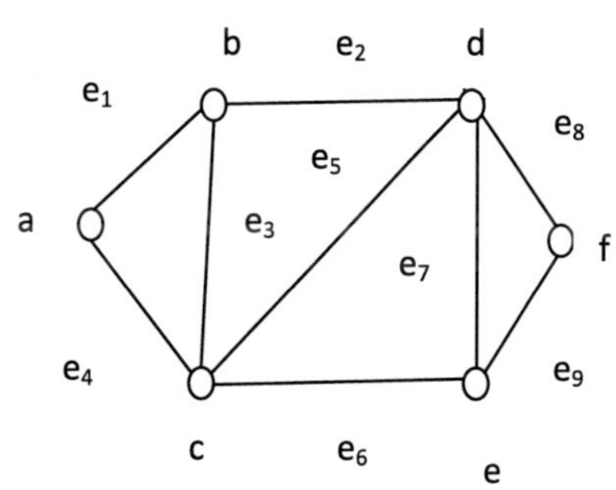

Figure 4.20

Solution

First choose a vertex say d (arbitrarily) as the root. Add the edges incident to this vertex d. Hence the edges e_2, e_5, e_7, e_8 are incident with the vertex d. these vertices create level 1 i.e.,

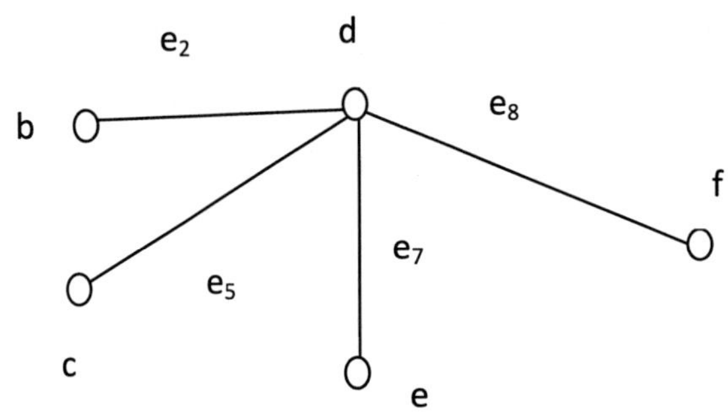

Figure 4.21

Now add edges which are incident to the vertices b, c, e, fin such a way that the resulting graph does not contain any circuit.

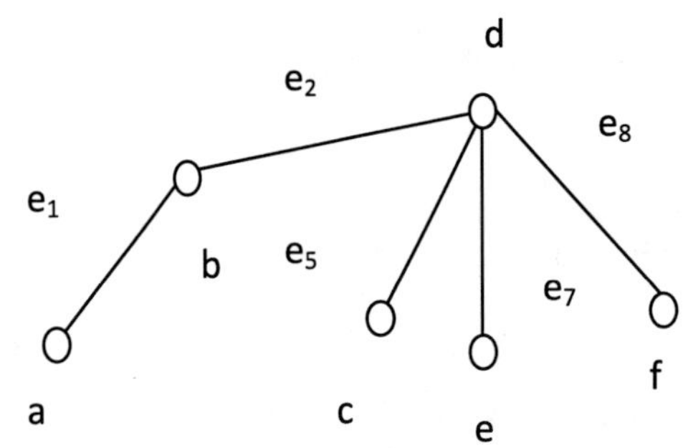

Figure 4.22

Thus at this level itself we have got the spanning tree T

Note: If the given graph is directed graph then we construct the underlying undirected graph and apply DF's or BF's to obtain spanning graph.

4.6.2 DEPTH FIRST SEARCH IN DIRECTED GRAPHS

We can modify both depth first search and Breath first search so that they can run given a directed graph as input. However, the output will not necessarily be a spanning tree, but rather a spanning forest. In both algorithms we can add an edge only when it is directed away from the vertex that is being visited and to a vertex not yet added. If at a stage of either algorithm we find that no edge exists starting at a vertex already added to one not yet added, the next vertex added by the algorithm becomes the root of a new tree in the spanning forest. This is illustrated in the example that follow.

Example 8:

Using DFS for the Directed graph of Graph of Figure 4.23. (a).

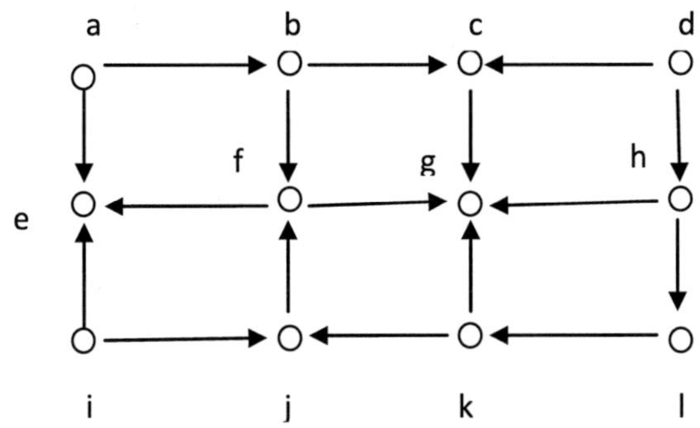

Figure 4.23(a)

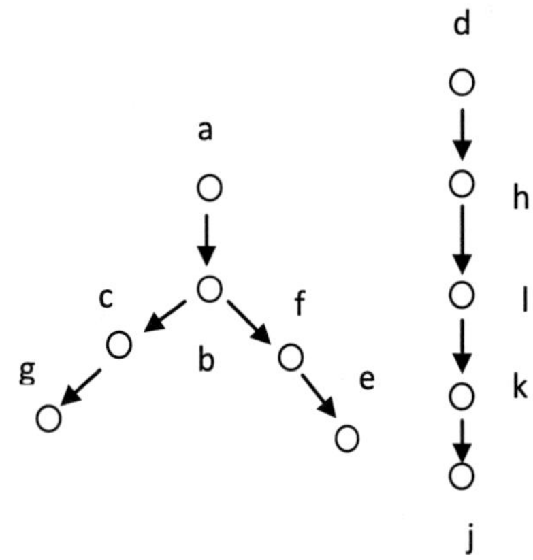

Figure 4.23(b)

In this example, we begin with the DFS at vertex a and add vertices b, c, and g and the corresponding edges where we are blocked. We backtrack to c but we are still blocked, and then backtrack to b, where we add vertices f and e and the corresponding edges. Backtracking takes us all the way back to a. We then start a new tree at d and add vertices h, l, k, and j and the corresponding edges. We backtrack to k, then l, then h, and back to d. Finally, we start a new tree at I, completing the DFS. the outputshown in Figure 4.23(b).

DFS in directed graphs is the basis of determining whether a directed graph has a circuit, a topological sort of a graph and it can also be used to find the strongly connected components of a Digraph. An important application of DFS and BFS in Digraphs, is to index websites using programmes called Web spiders (or crawlers or boots) by search engines like Google, Yahoo etc.

4.7 FUNDAMENTAL CIRCUIT AND CYCLIC INTERCHANGE

Let us recollect the definition of spanning tree. A sub graph of a connected graph is called as a spanning tree if the sub graph is a tree containing all the vertices of the graph. A connected graph might contain more than one spanning tree. Consider the following graphs. .

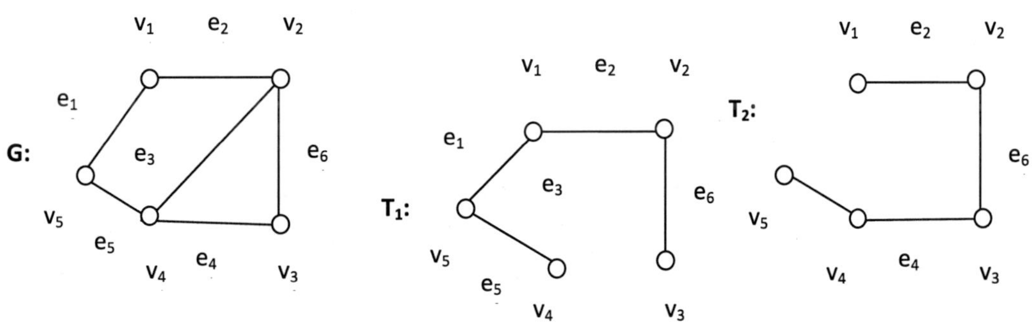

In T_1 the edges e_1, e_2, e_5, e_6 are present, whereas in T_2 the edges e_2, e_4, e_5, e_6 are present.

The edges of G which are present in a spanning tree T are called as the *branches* of G with respect to T. The edges of G which are not present in its spanning tree T are called the *chords* of G with respect to T.

In the above example, the branches of G are e_1, e_2, e_5, e_6 with respect to T_1: the branches of G are e_2, e_4, e_5, e_6 with respect to T_2.

Note: Let G be a connected graph on n vertices; e-edges and T be one of its spanning tree. Since T is a tree on n vertices, it has (n-1) edges. i.e., the number of branches of G with respect to t is (n-1); The number of chords of G with respect to T is e-(n-1). Often the number of branches of G is called as rank of G and is denoted by r(G); the number of chords of G is called as the nullity of G denoted by μ (G). In general, for a connected graph on n-vertices and e-edges, r(G), the rank of G is (n-1); μ (G), the nullity of G is e-n+1.

Fundamental Circuit: Let T be a spanning tree of a connected graph G, and e be a chord of G with respect to T. Since the spanning tree T is minimally acyclic, the graph T+e contains a unique cycle. This cycle is called a fundamental cycle in G with respect to T.

Every chord of G give rise to a fundamental cycle. Therefore, the number of fundamental cycles possible for a connected graph is almost μ(G).

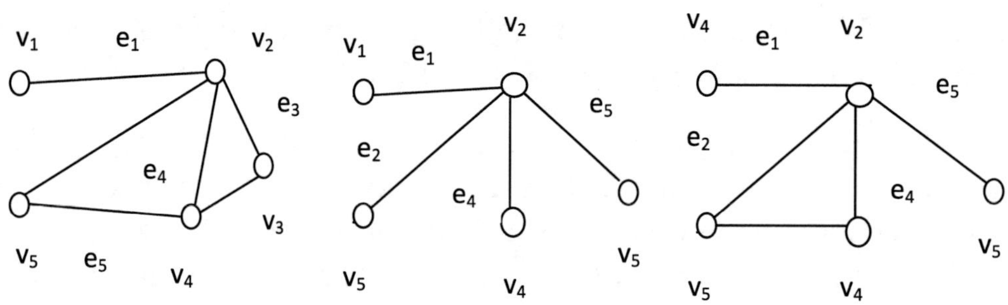

Figure 4.24(a) Graph G **(b)** Spanning tree of G **(c)** Fundamental circuit.

4.8 FINDING ALL SPANNING TREES OF A GRAPH

Usually, in a given connected graph there are a large number of spanning trees. In many applications we require all spanning trees. There is a procedure for obtaining additional spanning trees from a given spanning tree.

Cyclic Interchange: Let T be a spanning tree of G and e be a chord of G with respect to T. The graph T+e is a fundamental circuit. In this circuit other than edge e, all the other edges are branches of G with respect to T. On removal of any of the branches from the fundamental circuit, we get a spanning tree T_1 i.e., b is a branch in the fundamental circuit (with respect to the chord e_4), then spanning tree T_1 is obtained by removing b from T+e i.e., T_1= T+e-b. This process is called cyclic interchange.

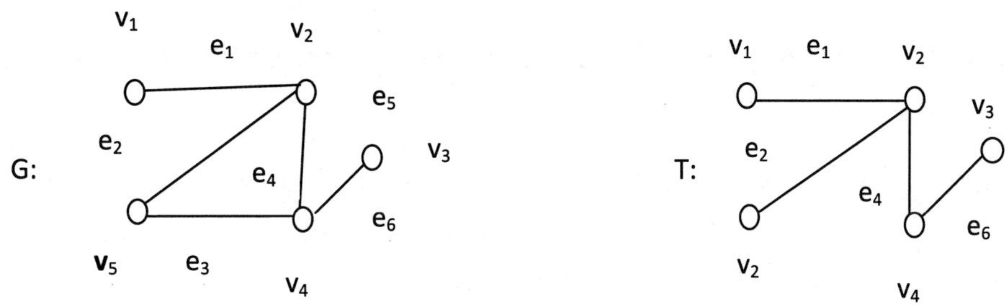

(a) A connected graph G (b) T-spanning tree. of G

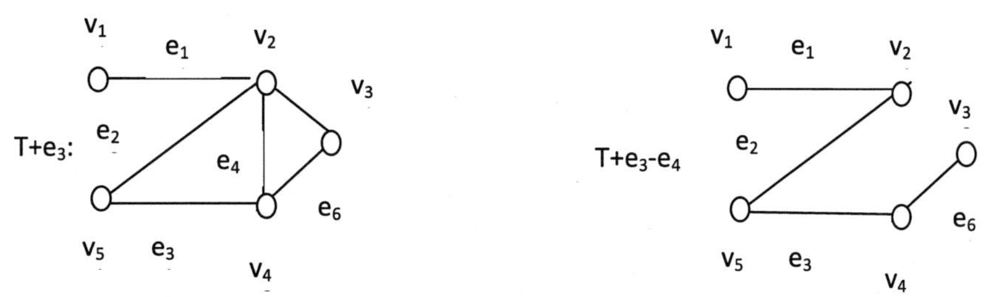

(c) T+$e_{3\text{-}}$ Fundamentalcircuit (d) T_1 – A spanning tree by cyclic interchange

Figure 4.25

Example 9:

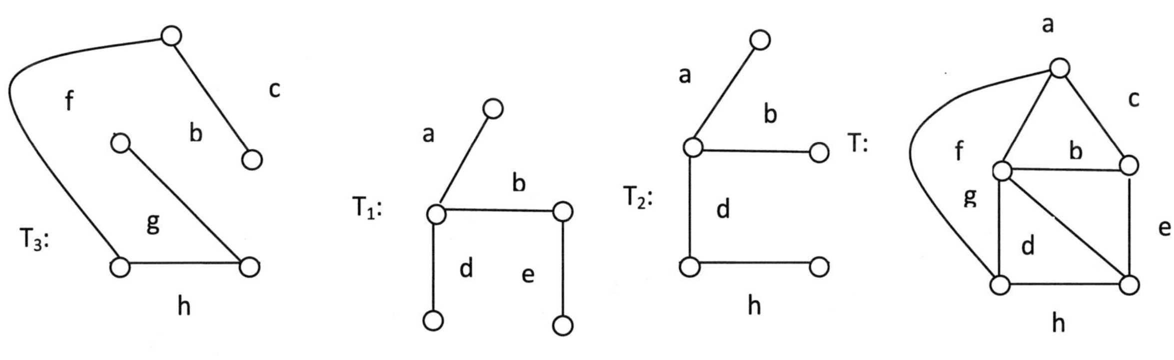

Figure 4.26

In Figure 4.26, T_2 and T_3 were obtained from T_1 by cyclic interchange or elementary tree transformation (addition of an edge and deletion of an appropriate branch) as follows: First edge d was added to T_1, giving the cycle abcd; removing a gives T_2 and removing b gives T_3. Removing c would give a fourth spanning tree, and, if we wanted more, we could return to T_1 and add an edge e, say, thereby completing another cycle abcef. Successively removing each of the edges a, b, c, and f of this cycle would give four new spanning trees. Thus we have a procedure for generating spanning trees for any given graph. Hence following theorem.

THEOREM 4. 6

Starting from any spanning graph G, we can obtain every spanning tree of G by successive cyclic exchanges.

The procedure outlined above raises many questions? Can we start from any spanning tree and get desired spanning tree by a number of cyclic exchanges? How long will have to continue exchange of edges? Out of all spanning trees is there a preferred one to start with? How many spanning trees does a connected graph possess? For this, there is a pretty result established in 1847 by the German Physicist Gustav Kirchhoff (1824 -1887), who was studying Electrical Networks. Spanning trees were used by Kirchhoff to generate a "Cycle Basis"for the cycles in the graphs of electrical networks.

4.9 KIRCHHOFF'S THEOREM

Let A be the adjacency matrix of a connected graph G and M be the matrix obtained from the adjacency matrix of a connected graph G by changing all 1's to -1 and each diagonal 0 to the degree of the corresponding vertex. Then the number of spanning trees of G is equal to the value of any co factor of M.

Example 10:

Consider the Graph G

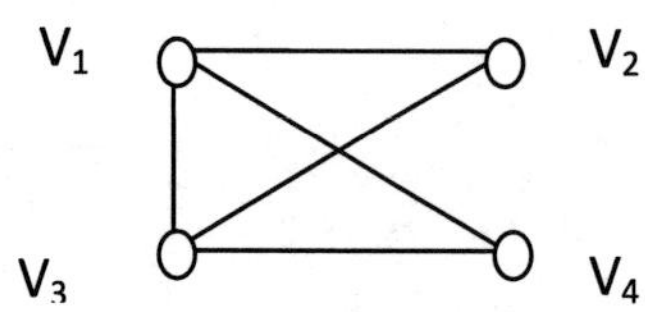

Figure 4.27

The adjacency matrix of the graph G is: $A = \begin{pmatrix} 0 & 1 & 1 & 1 \\ 1 & 0 & 1 & 1 \\ 1 & 1 & 0 & 0 \\ 1 & 1 & 0 & 0 \end{pmatrix}$

The matrix specified in Kirchhoff'sTheorem is

$$M = \begin{pmatrix} 3 & -1 & -1 & -1 \\ -1 & 3 & -1 & -1 \\ -1 & -1 & 2 & 0 \\ -1 & -1 & 0 & 2 \end{pmatrix}$$

The (1, 1) cofactor of M is

$$+\det\begin{bmatrix} 3 & -1 & -1 \\ -1 & 2 & 0 \\ -1 & 0 & 2 \end{bmatrix}\begin{bmatrix} 3 & -1 & -1 \\ -1 & 2 & 0 \\ -1 & 0 & 2 \end{bmatrix} = 3\begin{vmatrix} 2 & 0 \\ 0 & 2 \end{vmatrix}\begin{vmatrix} 2 & 0 \\ 0 & 2 \end{vmatrix} - (-1)\begin{vmatrix} -1 & 0 \\ -1 & 2 \end{vmatrix}\begin{vmatrix} -1 & 0 \\ -1 & 2 \end{vmatrix} + (-1)\begin{vmatrix} -1 & 2 \\ -1 & 0 \end{vmatrix}\begin{vmatrix} -1 & 2 \\ -1 & 0 \end{vmatrix}$$

$$= 3(4) + 1\,(-2) + (-1)\,2 = 8.$$

Kirchoff's theorem Guarantees that all the cofactors of m equal 8. For example, the (2, 3) co factor of M is

$$-\det\begin{bmatrix} 3 & -1 & -1 \\ -1 & -1 & 0 \\ -1 & 0 & 2 \end{bmatrix}\begin{bmatrix} 3 & -1 & -1 \\ -1 & -1 & 0 \\ -1 & 0 & 2 \end{bmatrix} = -[(-1)\,(0) + 2\,(-3-1)] = -(-8) = 8.$$

Thus, there are eight spanning trees for the graph G.

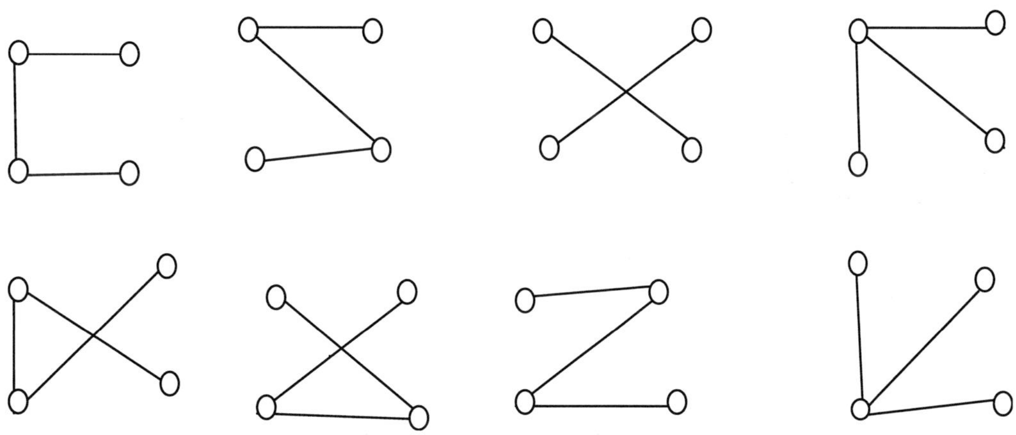

THEOREM 4.17

A non empty intersection of two fundamental circuits is always a path.

Proof: Let T_1 be a specified spanning tree of a graph G. Let e_1 and e_2 be two chords forming fundamental circuits C_1 and C_2 respectively. Then if $C_1 \cap C_2$ contains two edges u and v not connected in $C_1 \cap C_{2,}$ there is a path P_1 and C_1 between u and v; and this path does not contain a chord e_1. Similarly, there is a path P_2 in C_2 between u and v that does not contain chord e_2. Then the sub graph $P_1 \cup P_2 \cup \{u, v\}$ contains a circuit without containing any chord, which is impossible.

THEOREM 4.18

A set of edges is a cutest of a connected graph G if and only if S is minimal and contain at least one branch of every spanning tree of G.

Proof: Let S be a cut set of G. Let *t* be a spanning tree of G. Suppose S does not contain any branch of T. Then all the edges of T are present in G-S. Thus G-S is a connected graph, a contradiction to the fact that s is a cut set.

Conversely, Let S be the set of edges of G such that S contain at least one branch of every spanning tree, having minimum number of edges. Suppose G-S is connected, then we have G-S, contains a spanning tree T. Further, no edges of T is present in S, a contradiction. Since S is minimal G-S+e is a connected graph for all $e \in S$. Hence S is a cut set.

THEOREM 4.19

A cycle and the complement of any spanning tree must have at least one edge in common.

Proof: If there is circuit that has no common edge with the complement of a spanning tree, the circuit is contained in the spanning tree, which is not possible, since a tree cannot contain a circuit.

THEOREM 4.20

A cut set and any spanning tree must have at least one edge in common.

Proof: If there is cut set that has no common edge with a spanning tree, the removal of the cut set will leave the spanning tree intact. This means that the removal of the cut set will not separate the graph in to two components which is a contradiction to the definition of cut set.

THEOREM 4.21

Let T be a spanning tree and S be the fundamental cut set of G, determined by branch e with respect to T. Then the edge e contained in a fundamental circuit C of G with respect to T if and only if C is the fundamental circuit of G determined by a chord in S.

Proof: Let T be a spanning tree and S be the fundamental cut set of G, determined by branch e with respect to T. Then $S = \{ e, e_1, \ldots\ldots\ldots, e_k\}$, where $e_1, \ldots\ldots\ldots, e_k$ are the chords with respect to T. Let e_i be a chord of G with respect to T. Let C be the fundamental circuit of G determined by the chord e_i. Then $E(C) = \{ e_i, e_1', \ldots\ldots\ldots e_k' \}$ where $e_1', \ldots\ldots\ldots e_k'$ are branches of T. Let the branches $e \in E(C)$. Then $e \in E(C) \cap S$ and $e = e_j'$ for some j, $1 \leq j \leq k$. But the number of edges common to a cutest S and the circuit of G common to C and S other than e. However the only possible common edge in S and C other than e is a chord e_i. Thus $e_i \in S$. Thus, C is the fundamental circuit determined by the chord e_i belonging to S.

THEOREM 4.22

Let T be a spanning tree and C be the fundamental circuit set of G, determined by chord e with respect to T. Then the chord e contained in a fundamental cut set S of G if and only if s is the fundamental cut set of G determined by the branches in C.

Proof: Let T be a spanning treeand C be the fundamental circuit of G, determined by chord e with respect to T. Then $E(C) = \{e, e_1, e_2, \ldots\ldots\ldots, e_k\}$ where $e_1, e_2, \ldots\ldots\ldots, e_k$ are the branches with respect to T. Let e_i be a branch of T. Let S be the fundamental cut set of G determined by the branch e_i. Then $S = \{e_i, e_1', \ldots\ldots e_k'\}$ where $e_1', \ldots\ldots e_k'$ are chords of G with respect to T. Let the chord $e \in S$. Then $e \in E(C) \cap S$ and $e = e_j'$ for some j, $1 \leq j \leq k$. But the number of edges common to any cut set and a circuit of G is always even, implying that there exist one more edge of G common to C and S other than e. But the only possible common edge in S and C other than e is a branch e_i. Thus $e_i \in E(C)$. Thus S is the fundamental cutest determined by the branch e_i belonging to C.

4.10 CAYLEY'S FORMULAE

Cayley's was the first mathematician to obtain a formulae for the number of spanning of a labeled complete graph.

THEOREM 4.23 (CAYLEY'S)

The complete graph K_n, with $n \geq 3$ has n^{n-2} differentspanning trees.

Proof: After the vertices of the complete Graph K_n are labeled, any spanning tree of K_n is a labeled tree on n vertices. Conversely, any tree with these n vertices is a spanning treefor K_n because K_n contain all possible edges. Thus, the number of labeled trees on n vertices is the number of spanning trees for K_n, and the latter can be determined by Kirchhoff's Theorem. The adjacency matrix of K_n contains 0's on the diagonal and 1's everywhere the diagonal. So the matrix whose equal cofactors count the number of trees is the n×n matrix

$$M = \begin{pmatrix} n-1 & -1 & \cdots & -1 \\ -1 & n-1 & \cdots & -1 \\ -1 & -1 & \cdots & -1 \\ \vdots & \vdots & \ddots & \vdots \\ -1 & -1 & \cdots & -1 \\ 1 & -1 & \cdots & n-1 \end{pmatrix}$$

The (1, 1) cofactor of M is the determinant of an (n-1)×(n-1) matrix that looks exactly like M. Recall that the determinant of a matrix is not changed if a row is added to another row. Adding to the first row in turn each of the rows below it gives the matrix.

$$M = \begin{pmatrix} 1 & 1 & \dots & 1 \\ -1 & n-1 & \dots & -1 \\ -1 & -1 & \dots & -1 \\ \vdots & \vdots & \ddots & \vdots \\ -1 & -1 & \dots & -1 \\ -1 & -1 & \dots & n-1 \end{pmatrix}$$

And then adding the first row of this matrix to each of the rows below it give the matrix

$$\begin{pmatrix} 1 & 1 & 1 & \dots & 1 \\ 0 & n & 0 & \dots & 0 \\ 0 & n & & & 0 \\ \vdots & \vdots & \vdots & \ddots & \vdots \\ 0 & n & & & n \end{pmatrix}$$

Which is an upper triangular (n-1)×(n-1) matrix with n-2n's on the diagonal. Its determinant is n^{n-2}.

Tree Graph: The tree graph of a given graph G is defined as a graph in which each vertex corresponds to a spanning tree of G, and each edge corresponds to a *cyclic interchange* between the spanning tree of G represented by the two end vertices of the edge. The tree graph of any given (finite, connected) graph is connected.

4.11 SPANNING TREES IN A WEIGHTED GRAPH

A weighted graph is a connected graph G in which each edge has been assigned a real number w(e), called the weight of e. If H is a sub graph of a weighted graph, the weight w(H) of H is the sum of weights $w(e_1)+\dots\dots\dots+w(e_k)$, where $[e_1, \dots\dots\dots e_k]$ is the set of edges of H.

Many optimization problems could be solved by using weighted graphs. We try to find a sub graph of the weighted graph, with minimum (or maximum) weight. We look at a connected spanning sub graph of G which should not have a cycle. i.e., we look at a tree of G and finally we look for a spanning tree with minimum weights. Such a tree is called a **minimal spanning tree** or an optimal tree for G.

A **minimum spanning tree** of a weighted graph is a spanning tree of least weight, that is, a spanning tree for which the sum of the weights of all its edges is least among all spanning trees. Two algorithms due to Joseph Kruskal and Robert Prim are used to construct an optimal spanning tree. A necessary condition for these algorithms is that no weights should be negative. These algorithms are ***greedy***, because these help to make an optimal (minimal) choice at each step ignoring previous choices.

4.11.1 ALGORITHM 1: KRUSKAL'S ALGORITHM

The algorithm due to Kruskal, to construct a minimal spanning tree T for a connected Weighted graph Gwith n vertices andm edges is as follows: List all edges of the graph G in order of non decreasingweight. Next, choose an edge which has smallest weight among the edges of G which are not loops. Then for each successive step select (from all remaining edges of G) another smallest edge that makes no circuit with the previously selected edges. Continue until n-1 edges have been selected, and these edges will constitute the desired shortest spanning tree. This algorithm gives an acyclic sub graph T of G and the theorem given below proves that T is a minimal spanning tree of G. The steps of the algorithm are summarized below:

Step 1: Choose an edge of G, such that weight of e_1, $w(e_1)$ is as small as possible and e_1 is not a loop.

Step 2: If edges e_1, e_2, ...e_i have been chosen then choose an edge e_{i+1} Not already chosen, such that

(i) the induced sub graph G $[e_1, e_2, \ldots\ldots e_{i+1}]$ is acyclic and

(ii) w $(e_{i+1})$is as small as possible.

Step 3: If G has n vertices, stop after n-1 edges have been chosen. Otherwise repeat step 2.

THEOREM 4.17

Let G be a weighted connected graph in which the weights of the edges are all non-negative numbers. Let T be a sub graph of G obtained by Kruskal's algorithm. Then T is a minimum spanning tree of G.

Example 11:

Given below is a weighted graph with 5 vertices. Apply Kruskal's algorithm to find the minimal spanning tree.

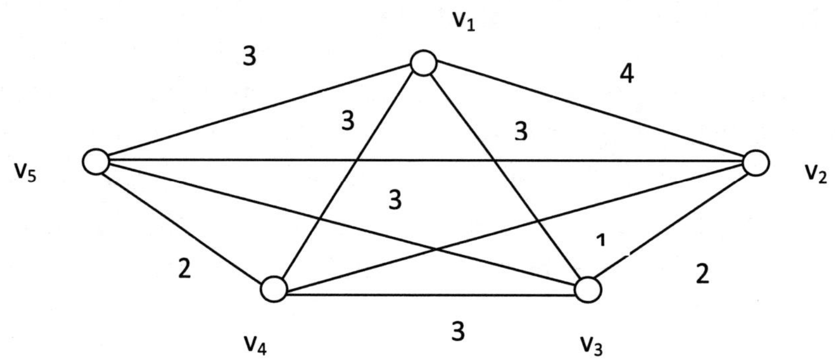

Solution

According to step 1, we will choose e = v_2v_4 as it has minimum weight.

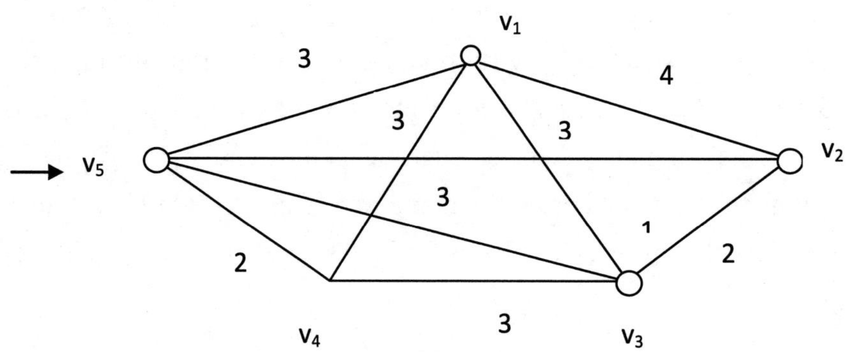

choose c = v_2v_3

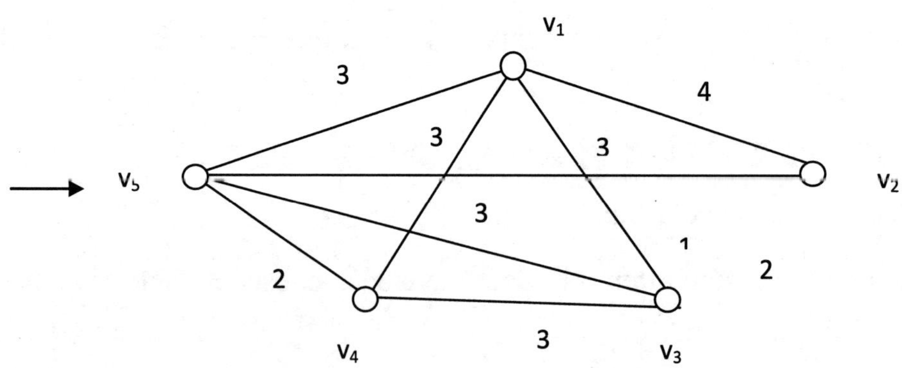

choose e = v_4v_5

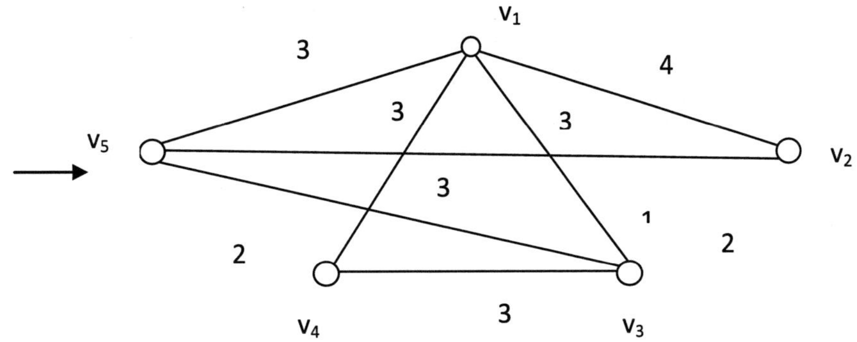

choose e = v_1v_4

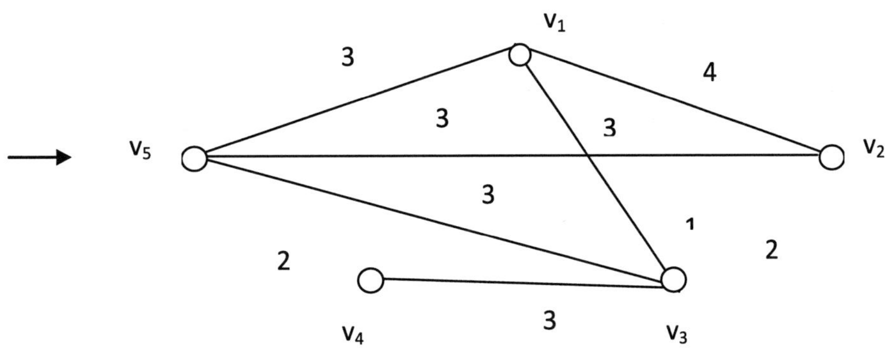

Since vertices are 5 and we have chosen 4 edges stop. The minimal spanning tree of the graph is

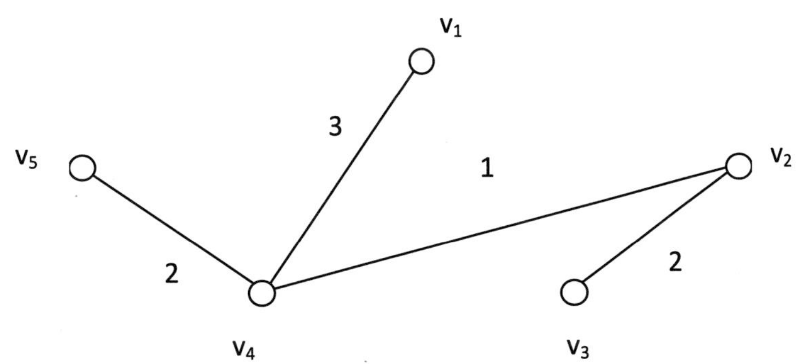

Note: Kruskal algorithm is an example of greedy algorithm. Greedy algorithms proceed by selecting the choice that looks best at the moment.

4.11.2 ALGORITHM: PRIM'S ALGORITHM

Another algorithm, which does not require listing all edges in order of non decreasing weight or checking at each step if a newly selected edge forms a circuit, is due to Prim. It chooses a vertex first and chooses an edge with smallest weight incident on that vertex. The algorithm involves following steps.

Step 1 : Choose any vertex v_1 of G.

Step 2 : Choose an edge $e_1 = v_1v_2$ of G such that $v_2 \neq v_1$ and e_1 has smallest Weight among the edges of G incident with v_1.

Step 3 : If edges $e_1, e_2, \ldots\ldots e_i$ have been chosen involving end points $v_1, v_2, \ldots\ldots v_{i+1}$, choose an edge $e_{i+1} = v_jv_k$ with $v_j \in [v_1, v_2, \ldots\ldots v_{i+1}$. such that e_{i+1} has smallest weight among the edges of G with precisely one end in $[v_1, v_2, \ldots\ldots v_{i+1}]$.

Step 4 : Stop after n-1 edges have been chosen. Otherwise go to step 3.

THEOREM 4.18

Let G be a weighted connected graph in which the weights of the edges are all non negative numbers. Let T be a sub graph of G obtained by Prim's Algorithm. Then T is a minimal spanning tree of G.

Example 12:

Find the minimal spanning tree of the weighted graph in previous Example using Prim's algorithm.

Solution

According to step 1 we choose vertex v_1. Now edge with smallest weight incident on v_1 is $e = v_1v_3$ so we choose $e = v_1v_3$ (or $e = v_1v_5$)

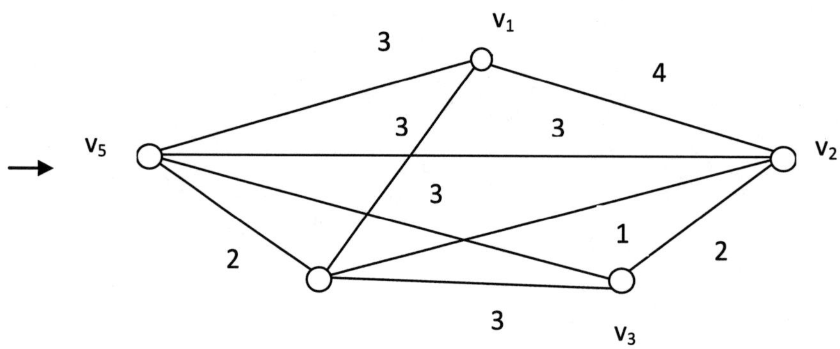

Choose $e = v_1v_3$. Now we look on the weights, $w(v_1, v_2) = 4$ $w(v_1, v_4) = 3$, $w(v_1, v_5) = 3$ $w(v_3, v_2) = 2$, $w(v_3, v_4) = 3$. Since minimum is $w(v_3, v_2)$, we choose $e = v_3v_2$.

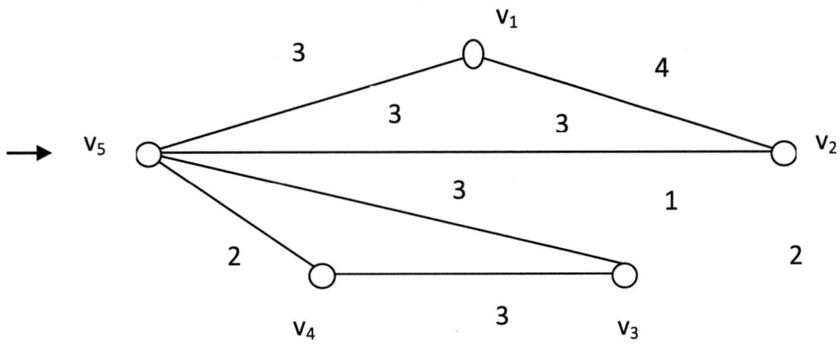

Again, w $(v_2, v_5) = 3$, w $(v_2, v_4) = 1$, and w $(v_3, v_4) = 3$.

Choose $e = v_2v_4$. Now, w $(v_4, v_5) = 2$

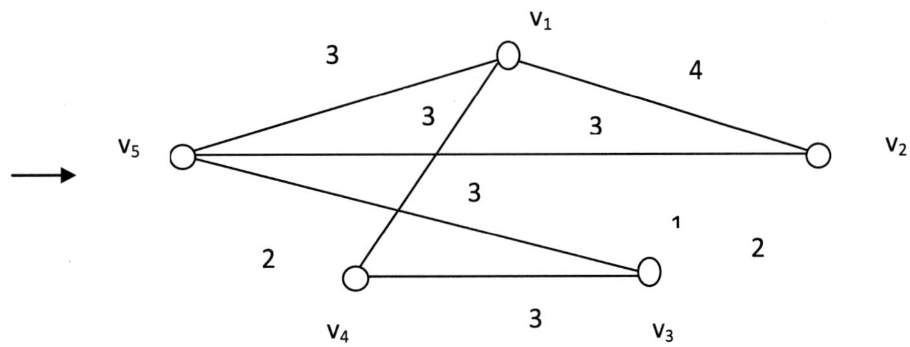

Choose $e = v_4, v_5$. The minimal spanning tree is

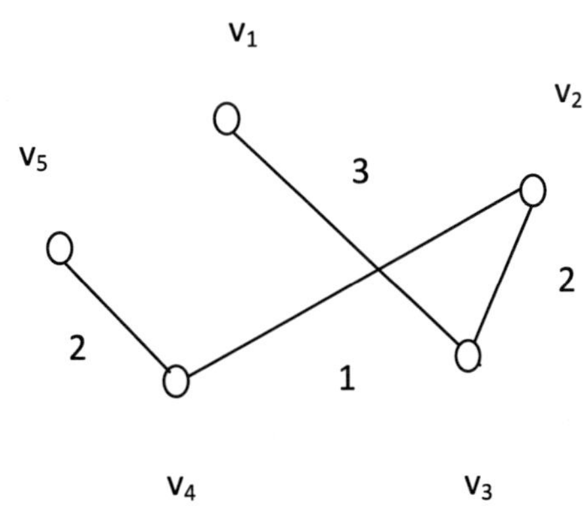

4.12 COMPUTER IMPLEMENTATION

A weighted graph is usually presented in matrix form. If G has n vertices and w_{ij} denotes the weight of an edge from vertex v_i to v_j, then we can present G as the n×n matrix with (i, j)[th] entry w_{ij} (subject to the assumption that G has no parallel edges). W_{ij} will be ∞ if there is no edge from v_i to v_j.

Example 12:

Represent the weighted graph given below in the matrix form.

Solution

The graph is presented in 6×6 matrix form.

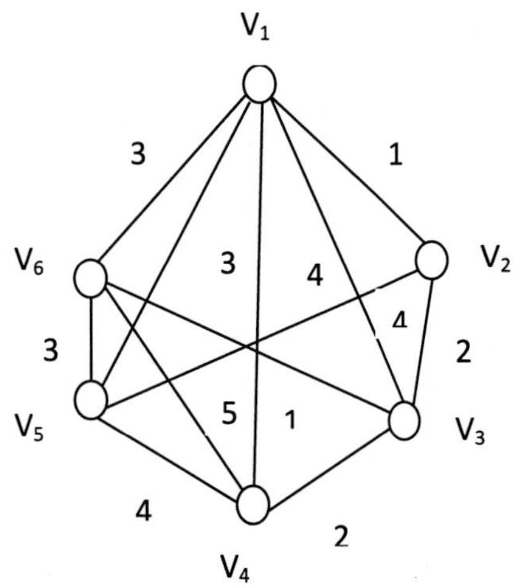

	v_1	v_2	v_3	v_4	v_5	v_6
v_1	∞	1	4	1	3	3
v_2	1	∞	2	∞	4	∞
v_3	4	2	∞	2	∞	∞
v_4	1	∞	2	∞	4	5
v_5	3	4	∞	4	∞	3
v_6	3	∞	∞	5	3	∞

Note that the diagonal is ∞, ∞, ..., ∞, and the matrix is symmetric with respect to the diagonal.

SUMMARY

1. A graph is called a tree if any two vertices in a given graph are connected by exactly one path. A **sub tree** of a tree is a sub graph that is also a tree. An undirected graph with no simple circuits is called a **forest**.

2. A **rooted tree** is a tree with a specified node, called **root** node (only one root in a tree), such that from this node, there is a unique path to any other node. The vertices adjacent to avertex v and below v are called **children** of v with v as their parent. A vertex that has no children is called a **leaf**. The **descendants** of a node v are its children all the way down from v. The **ancestors** of a node v are any node on the way from the path to v. A node that has children represents an **internal vertex**. The **height** of a rooted tree is the height of the lowest leaf. The height of a node, on the other hand, is the number of edges from the root to that node. A tree in which every leaf is at the same height is called a **balanced tree**.

3. A tree is called a **Binary tree** if each node has at most two children. A tree in which the children of each internal node are linearly ordered is called an **ordered tree**. The children of any node in a binary tree are ordered into a left and right child. The tree made up of a left child and all its descendants is called a **left sub tree. Right sub trees** are defined similarly. A **Binary Search tree** is a binary tree in which each internalnode v stores a data such that the data stored in the left sub tree of v are less than or equal to v and data stored in the right sub tree of v are greater than or equal to v.

4. **Infix** Notation is a form of an expression using parenthesis using the rule of operator precedence, with operators dispersed among the operands that connects the vertices of a the binary tree. In **prefix** notation the operator is written before the operands in the binary tree representing the expression whereas in **post fix** notation operands are placed before the operators in the binary tree representing the expression.

5. **Inorder traversal** of a binary trees is the listing of nodes of an ordered rooted tree defined recursively-by visiting the left sub tree first, followed by visiting the root node, followed by the right sub tree in the order they occur from left to right. **Preorder** tree traversal is the listing of nodes of an ordered rooted tree defined recursively-by visiting the root first followed by the left sub tree followed by the right sub tree in the order they occur from left to right. **Post order** traversal on the other hand, lists vertices of an ordered rooted tree defined recursively- by visiting the left sub tree first followed by right sub tree, followed by the root.

6. A **spanning tree** of a connected undirected graph G is a sub graph which is a tree and connects all vertices together. A single graph can have many spanning trees. A **minimum spanning tree** is a spanning tree with weight less than or equal to the weight of every other spanning tree.

7. **DFS** is a procedure of constructing a spanning tree, by adding edges that form a path until this is not possible, then backtracking the path until a vertex is found where a new path exists. **BFS** on the other hand is a procedure for obtaining a spanning tree, which successively adds alledges incident with the last set of edges added, until a simple circuit is formed.

8. **Prim's Algorithm** for producing a minimum spanning tree for a weighted graph, successively adds edges with minimum weight among all edges incident to a vertex that is already in the tree, such that addition of an edge does not produce a simple circuit.

9. **Kruskals Algorithm** for producing a minimum spanning tree for a weighted graph, successively addsedges of least eight that are not already in the tree, such that no edges produces a simple circuit when it is added.

EXERCISES

1. Check for the graph G whether it is (i) a tree (ii) Or a bridge?

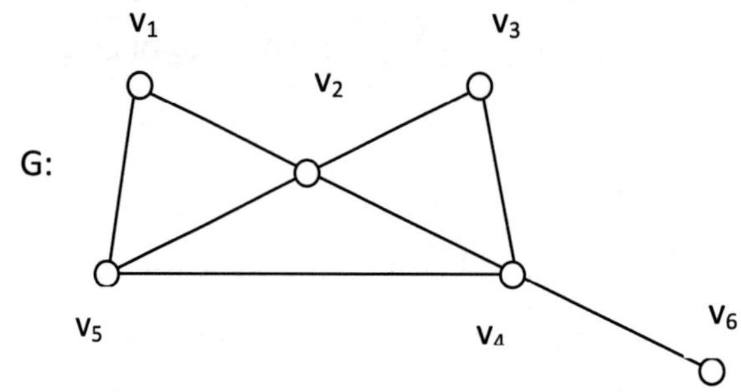

2. Use DFS and BFS to find the spanning tree of the graphG?

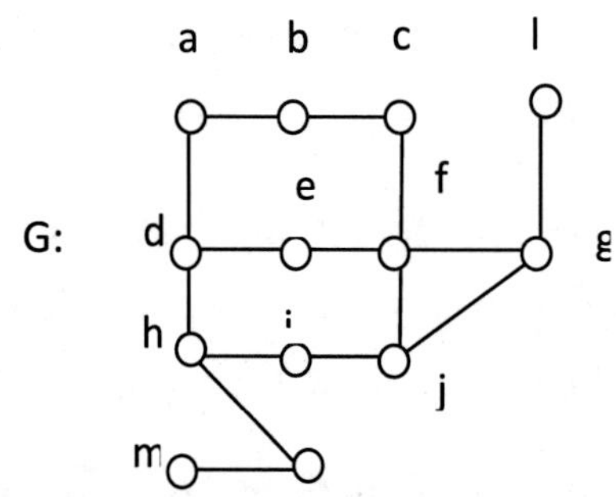

3. Use Kruskal's algorithm to find the minimal spanning tree of the followingconnected weighted graphs?

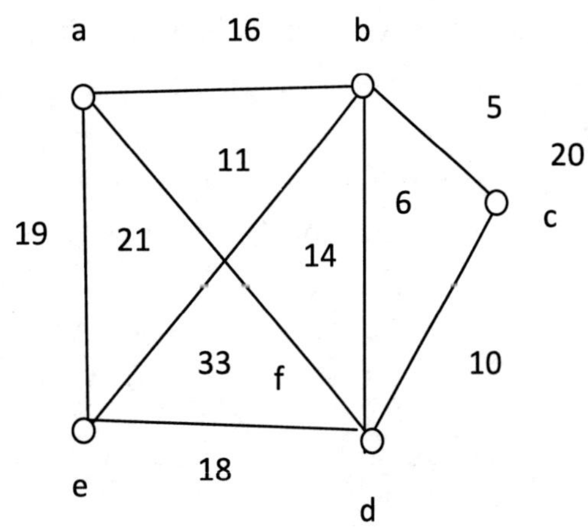

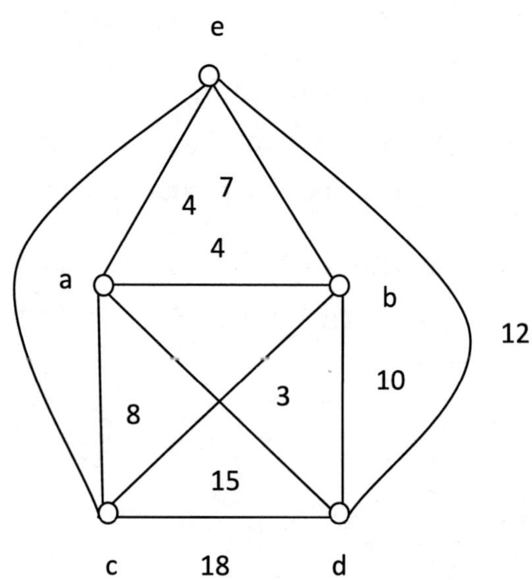

4. Write the expression ((X+2)3)*(y-(3 +x))-5 in (a) Prefix (b) postfix (c) infix notation

5. What is the value of the following prefix expression?-*33*425 (b) *+3+33+333.

6. Draw all spanning trees of the simple graph?

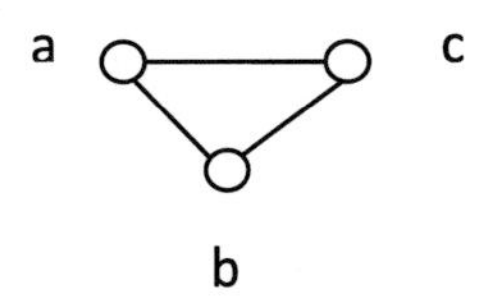

7. Use DFS to produce a spanning tree for the given simple graph?

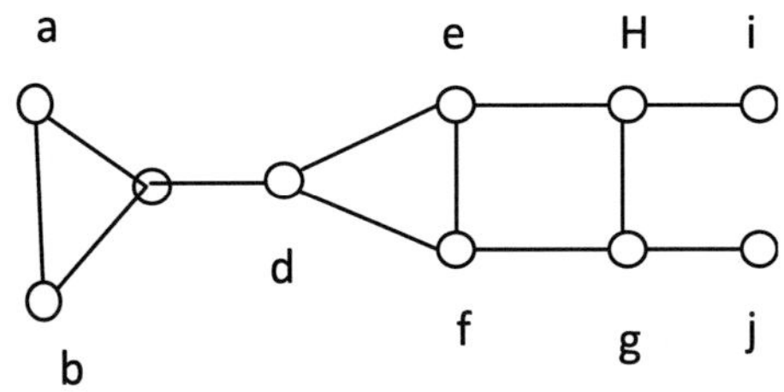

8. Draw the binary tree of the expression [((7-5) 2) 2] + (6(93))

9. Findthe MST of the weighted graph using Prim's algorithm?

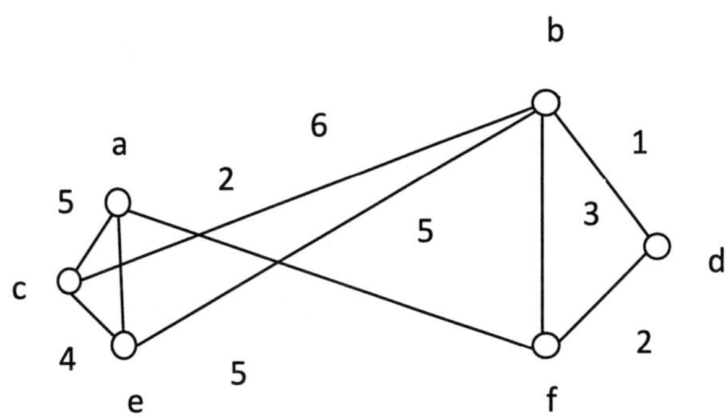

10. Find the MST of the weighted graph by Kruskal Algorithm?

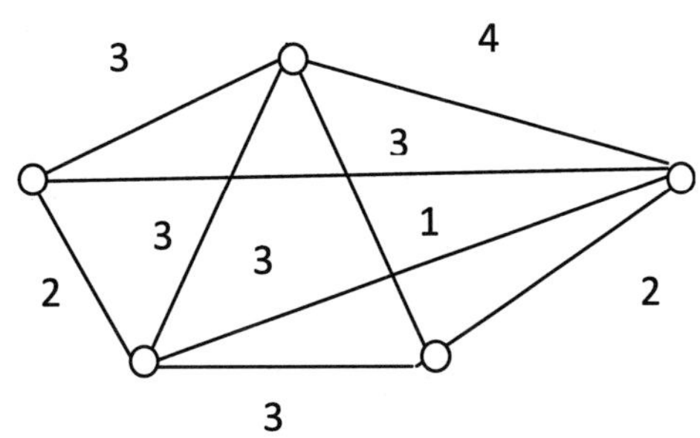

11. Write the(a) In order traversal (b) Preorder and (c) postorder traversal of the following trees T_1 & T_2?

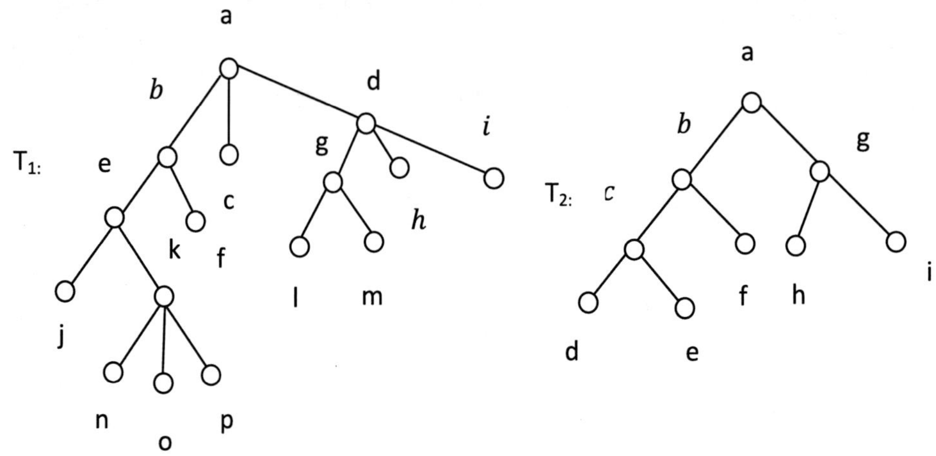

12. Apply DFS and BFS to produce a spanning tree for the following graph?

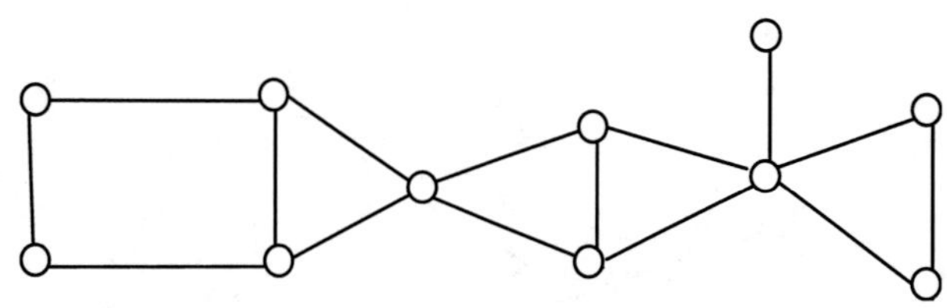

13. Represent the expression ((x+y))+((x-4)/3 as a binary tree and write the prefix and post fix forms of the expression?

14. Convert the following trees T_1& T_2 into Binary Trees?

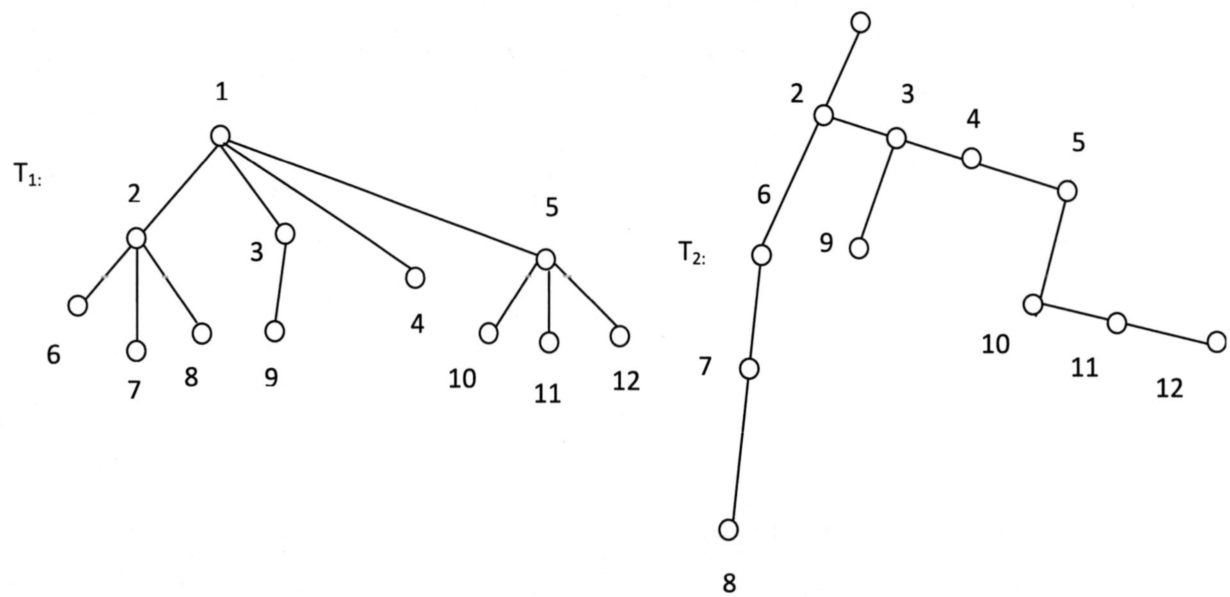

5

DISTANCE AND CENTRE

5.1 THE CENTER OF A GRAPH

In this section we define the concept of distance and eccentricity of a connected graph and extend the concept to define centre of a graph, as well as Digraph. .

For any two vertices u and v connected by a path in G, we define the distance between u and v, denoted by d (u, v), to be the length of a shortest u-v path. d (u, v) $= \infty$ ***if*** there is no path connecting them. A u-v path of length d(u, v) is called a u-v geodesic. In order for d(u, v) to be defined for all pairs u, v of vertices in G, the graph G must stay connected. We therefore assume that G is a connected graph. In any connected graph G, the term distance. satisfies all four of the following properties

1. d (u, v) ≥ 0 for all u, v $\in$ V(G).
2. d (u, v) = 0 if and only if u = v.
3. d (u, v) = d (v, u) for all u, v $\in$ V(G) (the symmetry property).
4. d(v, w) $\leq$ d(u, v) + d(v, w) for all u, v, w $\in$ V(G) (the triangle property).

The fact that the distance d satisfies properties 1-4 means that dis a metric and (V(G), d) is a metric space. It is ordinarily very useful when a distance is a metric as this concept has been studied widely. There are many concepts involving connected graphs that are defined in terms of distance which are valuable in providing information about these graphs.

Now, suppose, V be the vertex set of graph G then for each v $\in$ V, the **eccentricity** of v denoted by e(v), is the distance between v and a vertex farthest from v in G, defined by.

$$e(v) = \max\{ d(u,v): u \in v, u \neq v\}$$

The minimum eccentricity among the vertices of G is its **radius** and the maximum eccentricity is its **diameter**, which is denoted by **rad** (G), and **diam** (G) respectively. The radius denoted by rad (G) is defined by

$$\text{rad } G = \min (e(v): v \in V\}$$

While the diameter of G, denoted by diam G is defined by

$$\text{diam } G = \max \{e(v): v \in V\}$$

Or $$= \max\{d(u, v): u, v, \in V\}$$

A point v is a **central point** if $e(v) = \text{rad } G$, and the **center** of G is the set of all central points. If every vertex of G is a central vertex, then Cen (G) = G and G is called **self-centered.**

To illustrate the concepts we have just presented, consider the graph H of Figure 5.1. We will calculate eccentricity for one vertex, v_1 to illustrate the concept.

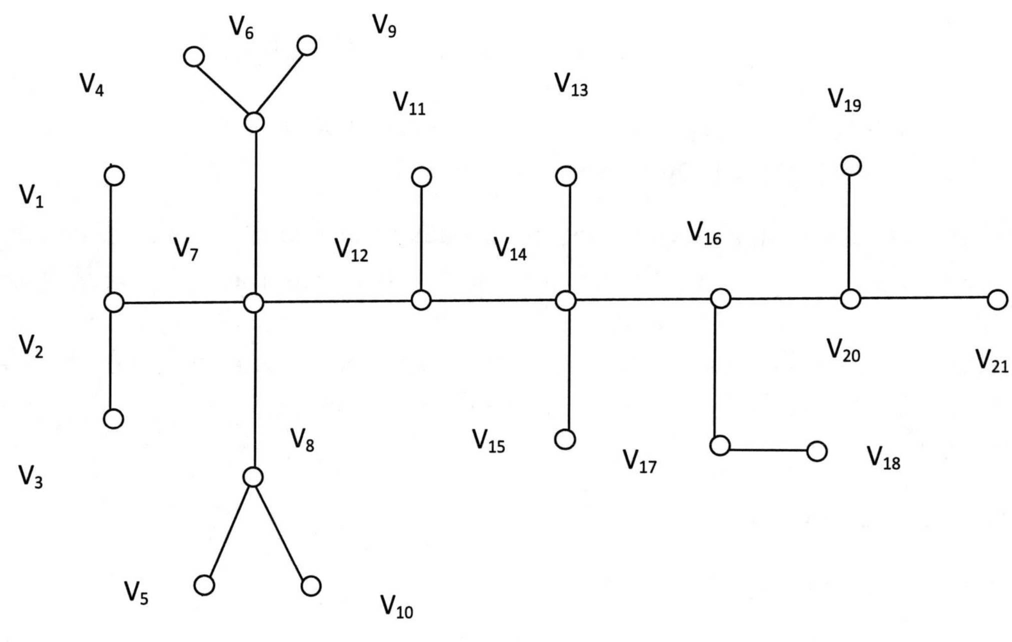

Figurc 5.1

d (v_1, u) where $u \in V$ where $V = (v_2, \ldots\ldots, v_{21})$ are:

$d(v_{1,} v_2) = 1$ $\quad d(v_1, v_7)= 2$ $\quad d(v_1, v_{12}) = 3$ $\quad d(v_1, v_{17}) = 6$

$d(v_1, v_3) = 2$ $\quad d(v_1, v_8)= 3$ $\quad d(v_{1,} v_{13}) = 5$ $\quad d(v_1, v_{18)}= 7$

$d(v_1, v_4) = 4$ $\quad d(v_1, v_9)= 4$ $\quad d(v_1, v_{14}) = 4$ $\quad d(v_1, v_{19}) = 7$

$d(v_{1,} v_5) = 4$ $\quad d(v_1, v_{10}) = 4$ $\quad d(v_{1,} v_{15}) = 5$ $\quad d(v_1, v_{20}) = 6$

$d(v_1, v_6) = 3$ $\quad d(v_1, v_{11}) = 4$ $\quad d(v_1, v_{16}) =5$ $\quad d(v_1, v_{21}) = 7$

So, **eccentricity** $e(v_1) = \max[1, 2, 3, 4, 5, 6, 7] = 7$.

In the similar way,

$e(v_2) = 6$	$e(v_7) = 5$	$e(v_{12}) = 4$	$e(v_{17}) = 6$
$e(v_3) = 7$	$e(v_8) = 6$	$e(v_{13}) = 5$	$e(v_{18}) = 7$
$e(v_4) = 7$	$e(v_9) = 7$	$e(v_{14}) = 4$	$e(v_{19}) = 7$
$e(v_5) = 7$	$e(v_{10}) = 7$	$e(v_{15}) = 5$	$e(v_{20}) = 6$
$e(v_6) = 6$	$e(v_{11}) = 5$	$e(v_{16}) = 5$	$e(v_{21}) = 7$

Since the smallest eccentricity is 4, rad (H) =4.

That is, **radius** = min [e (v)] = min [4, 5, 6, 7] = 4.

Because the largest eccentricity is 7, diam (H) = 7.

That is, **diameter** = max [e (v)] = max [4, 5, 6, 7] = 7

Since, vertices v_{14} and v_{12} have minimum eccentricity hence,

Centre = [v_{12}, v_{14}].

Example 1:

Find the eccentricities, radius and centre of the graph of Figure 5.2?

Solution

For the graph of Figure 5.2,

we have e (q) = e(u) = e (y) = e (s) = e (t) = e(x) = 3.

e (w) = e (v) = 4, e (r) = e (z) = 2.

Since the smallest eccentricity is 2, rad (H) = 2.

That is, **radius** = min [e (v)] = min [2, 3, 4] = 2.

Because the largest eccentricity is 4, diam (H) = 4.

That is, **diameter** = max [e (v)] = max [2, 3, 4] = 4.

Since, vertices r and s have minimum eccentricity hence,

Centre= [r, z].

The centre of H is also showninseparately in Figure 5.2.

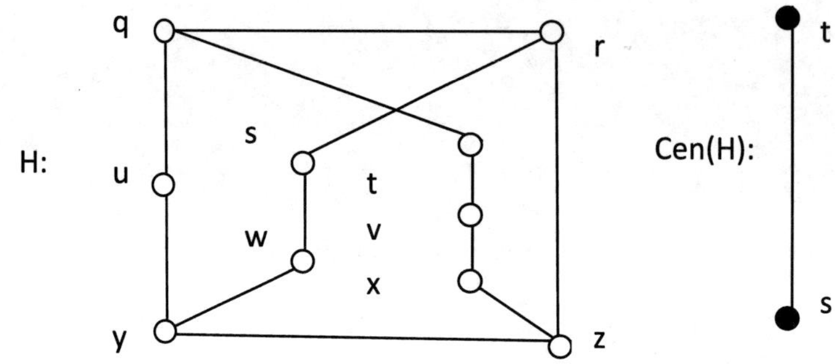

Figure 5.2

There are a number of observations that can be made about the graph H of Figure 5.3. We have already mentioned that rad (H) = 2 and diam (H) = 4. It is an established fact in geometry, where, of course, the diameter is always twice the radius . This fact together with, the knowledge that diam (H) = 2 rad (H) for the graph H in Figure 5.2 might reasonably suggest that diam (G) = 2 rad (G) for every connected graph V. However, this is not the case for all graphs G, as the graph 5.3 shows. In the three graphs G_1, G_2, and G_3, of Figure 5.3, each graph has radius 2, where diam (G_k) = k for k = 2, 3, 4. That is, Figure 5.3 illustrates the only possible diameters for a graph having radius 2. Therefore, there is no identity that relates the radius and the diameter of a graph.

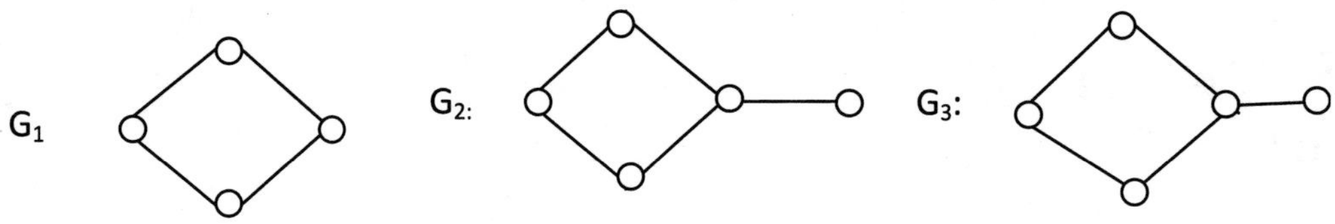

Figure 5.3

THEOREM 5.1

For every nontrivial connected graph G,

$$\text{rad}(G) \le \text{diam}(G) \le 2\,\text{rad}(G).$$

Proof: The inequality rad (G) ≤ diam (G) holdssince the smallest eccentricity cannot exceed the largest eccentricity. Let u and v be two vertices such that d (u, v) = diam (G) and let w be a central vertex of G. Therefore, e(w) = rad (G). Hence the distance between w and any other vertex of G is at most rad (G). By the triangle inequality,

$$\text{diam}(G) = d(u, v) \le d(u, w) + d(w, v) \le \text{rad}(G) + \text{rad}(G) = 2\,\text{rad}(G).$$

Another observation about the graph H in Figure 5.2. is that the eccentricities of every two adjacent vertices differ by at most 1 and is true for all connected graphs.

THEOREM 5.2

For every two adjacent vertices u and v in a connected graph,

$$|e\ (u) - e\ (v)| \leq 1$$

Proof: Assume, without loss of generality, that e (u) ≥ e (v). Let x be a vertex that is farthest from u. So d (u, x) = e (u). By the triangle inequality,

$$e\ (u) \geq d\ (u, x) \leq d\ (u, v) + d\ (v, x) \leq 1 + e\ (v)$$

Hence e (u) ≤ 1 + e (v), which implies that 0 ≤ e (u) – e (v) ≤ 1. Therefore,

$$|e\ (u) - e\ (v)| \leq 1$$

In the same way the following can be proved.

THEOREM 5.3

Let u and v be adjacent vertices in a connected graph G. Then

$$|d\ (u, x) - d\ (v, x)| \leq 1$$

Returning once again to the graph H of Figure 5.1, we see that Cen (H) ≅ K_2. This brings up a natural question. Which graphs can be the center of some graph? In an observation made by Stephen Hedetniemiin his works, showed that "every graph" is the answer to this question.

THEOREM 5.4

Every graph is the center of some graph.

Proof: Let G be a graph. First, construct a graph H from G by adding two new vertices u and v to G and join them to every vertex of G but not to each other. Next we add two other vertices u_1 and v_1 to H, where we join u_1 to u and join v_1 to v. The resulting graph H, is shown in Figure 5.4.

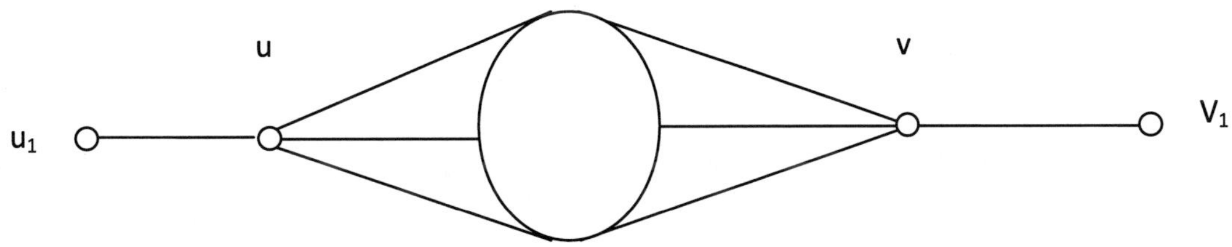

Figure 5.4

Since e (u_1) = e (v_1) = 4, e (u) = e (v) = 3, and e (x) = 2 for every vertex x in G, it follows that V(G) is the set of central vertices of H and so Cen (H) = G.

5.2 DISTANT VERTICES

A vertex v in a connected graph G is called a **peripheral vertex** if e (v) = diam (G). Thus, in certain sense, a peripheral vertex is opposite to a central vertex. The sub graph of G induced by its peripheral vertices is the **periphery** per (G).

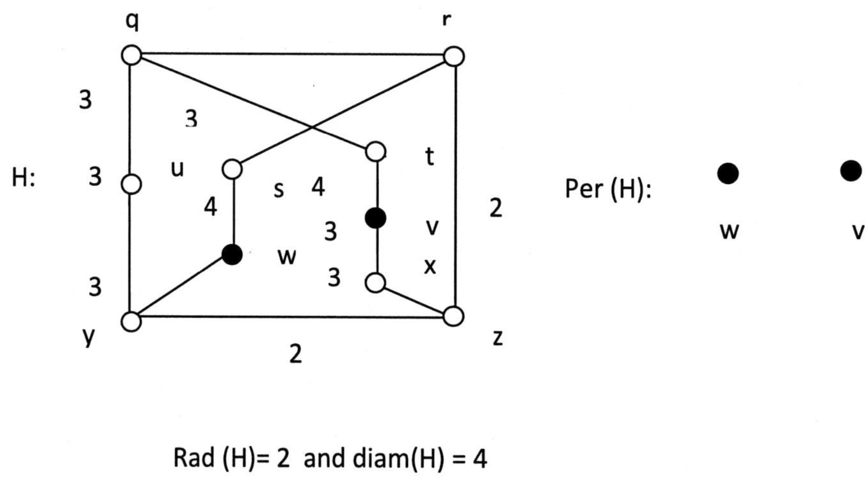

Figure 5.5 Maximal planar graphs

For the graph H of Figure 5.2, the periphery of H is shown in Figure 5.5, and is isomorphic to $2K_1$. That is, it consists of two isolated vertices and so it is disconnected. A natural question that may ariseat this point is whether periphery of every graph is disconnected? The answer to this question is No, as the graph F of Figure 5.6 shows.

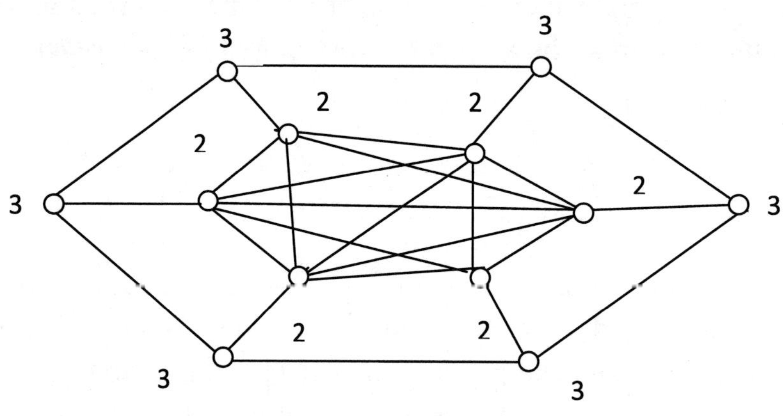

Figure 5.6 A Graph F

Here each vertex is labeled with its eccentricity and diam (F) = 3. Since diam (F) = 3, it follows that per (F) ≅ $C_{6,}$ which is connected. In fact, if G ≅ C_{n}, then per (G) ≅ C_n for n ≥ 3. Could it be then that every graph is the periphery of some graph? The answer is No. as the Theorem 5.6 below shows.

THEOREM 5.5

A nontrivial graph G is the periphery of some graph if and only if every vertex of G has eccentricity 1 or no vertex of G has eccentricity 1.

Proof: Assume first that every vertex of G has eccentricity 1. If every vertex of G has eccentricity 1, then G is complete and Per (G) = G. Next assume that no vertex of G has eccentricity 1. This implies that for every vertex u of G, there is a vertex v in G that is not adjacent to u. Let H be the graph obtained by adding a new vertex w and joining w to every vertex of G, Then $e_H(w) = 1$. Since $e_H(x) = 2$ for every vertex of G, Therefore, it follows that every vertex of G is a peripheral vertex of H and so Per (H) =G.

A vertex v is an eccentric vertex ofthe graph G, if v is farthest from some vertex of G. Every peripheral vertex of G is an eccentric vertex. But the converse is not true.

On the other hand, if every vertex of some graph G has the same eccentricity (and is therefore a peripheral vertex), then certainly every vertex of G is an eccentric vertex. However, the graph H of

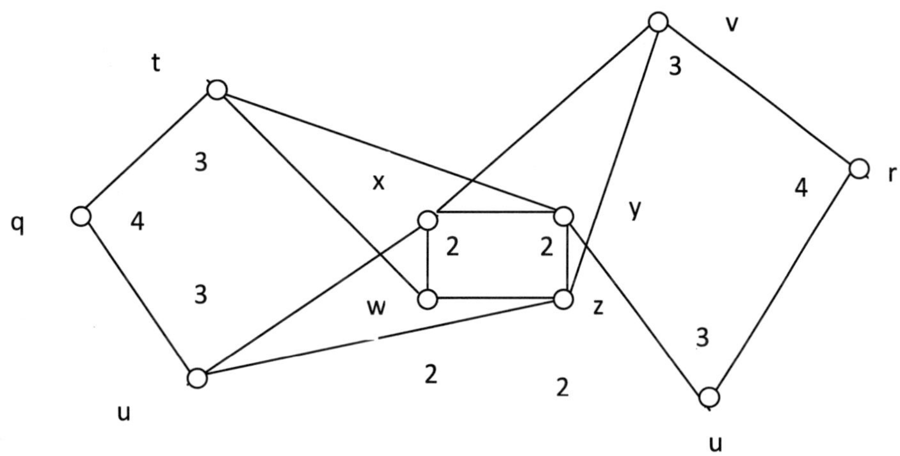

Figure 5.7

Figure 5.7 shows that every vertex of a graph can be an eccentric vertex without all the eccentricities being the same. A connected graph G is an eccentric graph if every vertex of G is an eccentric vertex. Therefore, the graph H of Figure 5.7 is an eccentric graph, as every vertices have the same eccentricity. Ordinarily, however, only some of the vertices of a graph are eccentric.

The **eccentric sub** graph Ecc (G) of a connected graph G is the sub graph of G induced by the set of eccentric vertices of G. If every vertex of a graph G is an eccentric vertex, then Ecc(G) = G. A connected graph F and its eccentric sub graph are shown in Figure 5.8. In the graph F of Figure 5.8, Ecc(G) $\cong 2\ P_3$.

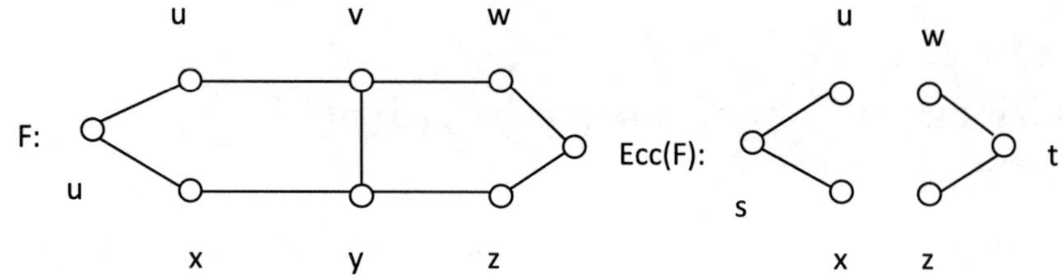

Figure 5.8

A natural question that one may ask at this point is, which graphs are eccentric sub graphs of some graph. Or which graphs are the **peripheries** of some graph? An answer to this brings up the following theorem.

THEOREM 5.6

A nontrivial graph G is the eccentric sub graph of some graph if and only if every vertex of G has eccentricity 1 or no vertex of G has eccentricity 1.

Proof: Assume that every vertex of a graph G has eccentricity 1 or no vertex of G has eccentricity 1. If every vertex of G has eccentricity 1, then G is complete and G is an eccentric graph. Thus, Ecc(G) = G. Next, assume that no vertex of G has eccentricity 1. Let H be the graph obtained by adding a new vertex w and joining w to every vertex of G. Since every vertex v in H that belongs to G is an eccentric vertex of w but w is not an eccentric vertex of any vertex of H, it follows that Ecc(H) = G.

If v is an eccentric vertex of a vertex u in a connected graph G, then no vertex of G is farther from u than v is. In particular, if w is a neighbor of v, then $d(u, w) \leq d(u, v)$. However, a vertex can have this property without being an eccentric vertex of u.

A vertex v in a connected graph G is a **boundary of a vertex** u if $d(u, w) \leq d(u, v)$ for each neighbor w of v; while a vertex v is a **boundary vertex of the graph** G if v is a boundary vertex of some vertex of G. A vertex v is a boundary vertex of u if no u–v geodesic can be extended at v to a longer geodesic. Intuitively, beginning at u, a boundary vertex of u is reached when, locally, it is not possible to proceed farther from u.

There are certain vertices in a nontrivial connected graph that cannot be boundary vertices.

THEOREM 5.7

No cut–vertices of a connected graph G is a boundary vertex of G.

Proof: Assume that there exists a connected graph G and a cut–vertex v of G such that v is a boundary vertex of some vertex u in G. Let G_1 be component of G–v that contains u and let G_2 be another component of G–v. If w is a neighbor of v that belongs to G_2, then $d(w, u) = d(u, v) + 1$, which contradicts the assumption that v is a boundary vertex of u.

Since no cut–vertex can be a boundary vertex, no cut-vertex can be an eccentric vertex or a peripheral vertex either. There are certain vertices, however, that must be boundary vertices.

A vertex v in a graph G is called a **complete vertex** (or an extreme vertex)if the sub graph of G induced by the neighbors of v is complete. In particular, every end vertex is complete. Therefore, if v is a complete vertex and u is a neighbor of v, then d(w, u) = d(w, v) = 1 for every w $\in$ N(v). Thus, v is a boundary vertex of u. a complete vertex v is not a boundary vertex of each neighbor of v, it is a boundary vertex of every vertex different from v.

There are certain vertices in a connected graph G that have a close connection with boundary vertices. Let x and z be two distinct vertices in G. A vertex y distinct from x and z is said to lie **between** x and z if

$$d(x, z) = d(x, y) + d(y, z)$$

that is, triangle inequality becomes an equality. A vertex v is an **interior vertex** of G if for every vertex u distinct from v, there exists a vertex w such that v lies between u and t. The interior Int(G) is the sub graph of G induced by interior vertices. For example, for the graph G of Figure 5.9, the vertices s, v, and x are the interior vertices of G and so Int (G) P_3. the interior vertices are precisely those vertices that are not boundary vertices.

5.3 DIRECTED DISTANCE

We now turn to digraphs. Let D be a connected digraph. For two vertices u and v of D, the directed distance d(u, v) is the length of a shortest u–v path in D. A u–v path of length d(u, v) is a u–v geodesic. The fact that D is connected does not guarantee the existence of a u-v path in D. For this reason, when discussing distance in digraphs, we ordinarily assume that D is strong (Typically as a strong oriented graph). Directed distance does, however, satisfy the triangle inequality

$$d(v, w) \leq d(u, v) + d(v, w) \text{ for all } u, v, w \in V(G)$$

For directed distance, the eccentricities of the vertices of a strong digraph are synonymous with the term eccentricity of graphs. Let D be a strong digraph and let v be a vertex of D. The eccentricity e (v) of v is the length of a longest geodesic from v to a vertex of D. The minimum eccentricity among the vertices of D is its radius rad(D) and maximum eccentricity is its diameter diam(D). In particular the diameter of D can be more than twice its radius and so for the directed distance there are no analogous theorems as in the case of distance in undirected graphs. .

SUMMARY

1. The **eccentricity** of a vertex v, denoted by e (v), is the distance between v and a vertex farthest from v in G, and is defined by. e(v) = max{d (u, v): u e v, u ≠ v}. The minimum eccentricity among the vertices of G is its **radius** and the maximum eccentricity is its **diameter**, which is denoted by **rad** (G), and **diam** (G) respectively. For directed distance, the eccentricities of the vertices of a strong digraph are synonymous with the term eccentricity of graphs.

2. A point v is a **central point** if e(v) = rad G, and the **center** of G is the set of all central points. If every vertex of G is a central vertex, then Cen (G) = G and G is called **self-centered.** A vertex v in a connected graph G is called a **peripheral vertex** if e (v) = diam (G). That is, a peripheral vertex is opposite to a central vertex. A vertex v is an **eccentric vertex of**the graph G, if v is farthest from some vertex of G. Every peripheral vertex of G is an eccentric vertex.

3. A vertex v in a connected graph G is a **boundary of a vertex**u if d (u, w) ≤ d (u, v) for each neighbor w of v; while a vertex v is a **boundary vertex of the graph** G if v is a boundary vertex of some vertex of G.

4. A vertex v in a graph G is called a **complete vertex** (or an extreme vertex)if the sub graph of G induced by the neighbors of v is complete. In particular, every end vertex is complete.

5. A vertex v is an **interior vertex** of G if for every vertex u distinct from v, there exists a vertex w such that v lies between u and t. The **interior** Int (G)is the sub graph of G induced by interior vertices.

EXCERCISES

1. For the graph shown below, findthe eccentricity ofall the vertices, radius and centre of the graph given below?

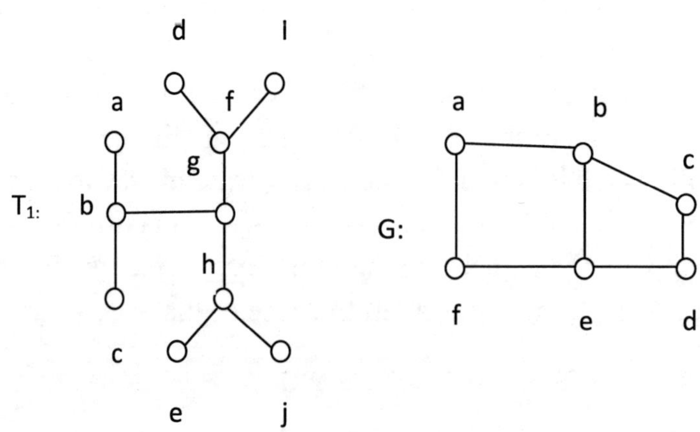

2. Find the distance of each of the vertices of the graphs given below:

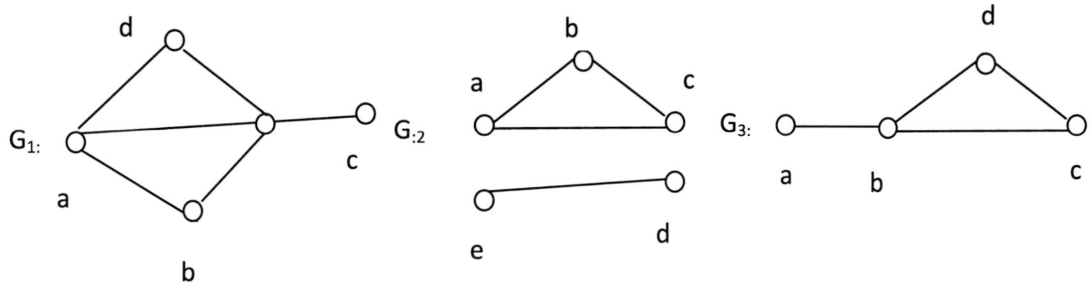

3. Find the radius and diameter of path graph for $n \geq 3$ P_n (for $n \geq 3$) & Cycle graph (for for $n \geq 3$) C_3?

4. Find the radius and diameterof Petersen Graph. ? What is its centre?

5. What is the periphery of P_n (for $n \geq 2$) and Petersen graph?

6. For the graph G, given below, determine

   (a) the set of peripheral vertices of G,

   (b) the set of eccentric vertices of G,

   (c) the set of boundary vertices of G,

   (d) the periphery and boundary of G?.

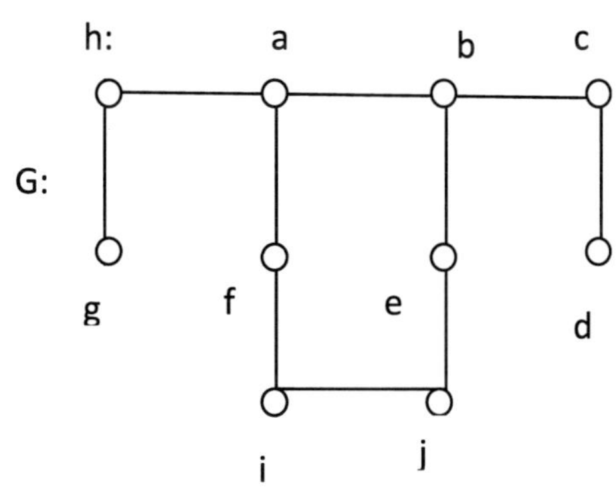

6

CONNECTIVITY

6.0 INTRODUCTION

It is probably clear by now that one of the most important properties that a graph can possess is that of being connected. The connectivity (a measure of connectedness) of a graph is a particularly intuitive area of graph theory and extends the concept of cut point, and bridge, Some connected graphs are connected rather loosely in the sense that deletion of an edge or of a vertex from the graph destroys connectedness of the graph. There are graphs in the other extreme as well, such as the complete graphs K_n, $n \geq 2$, which remains connected even after removal of all but one vertex. This suggests the possibility of introducing a measure of how connected a graph is, (i.e., the study of two graph parameters the edge and vertex connectivity) which is a goal of this section.

Figure 6.1 shows seven graphs of order 7. Obviously, all of these graphs are connected. The graph G_1 is a tree. $G_4 \cong C_7$, and $G_7 \cong k_7$. We now turn to connected graphs containing an edge (or vertex) whose removal results in a disconnected graph.

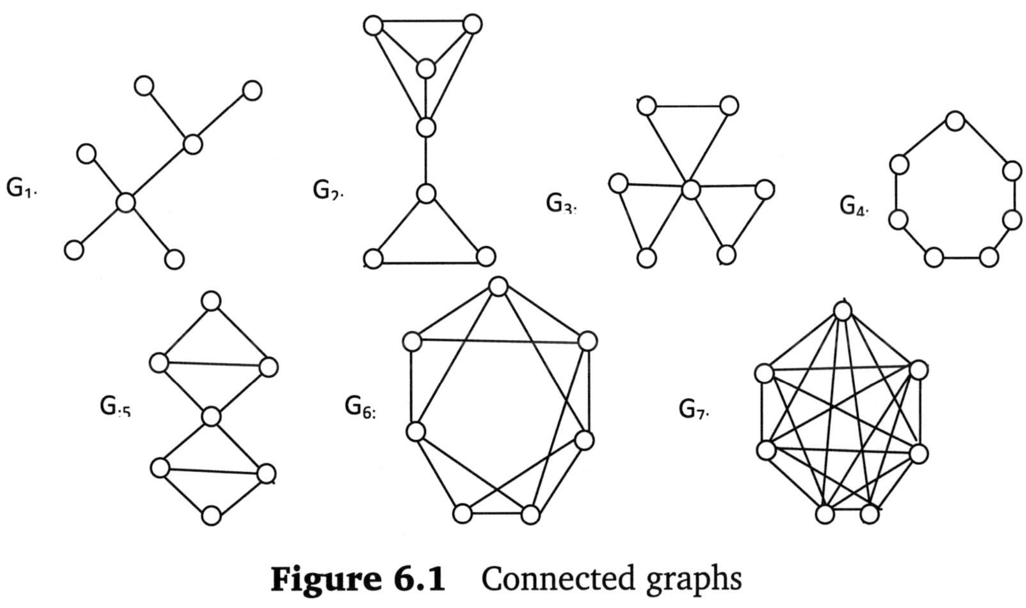

Figure 6.1 Connected graphs

6.1 CUT–SETS AND CUT VERTICES

Let G be a connected graph. If G contains an edge such that G-e is disconnected, then e is a **cut edge or bridge** of G. The graph $G_2$of Figure 5.1 has a bridge. So does G_1. In fact, every edge of G_1 is a bridge since G_1 is a tree. Further if G contains a vertex such that G–v is disconnected, then v is a **cut-vertex** of G. In fact, except for the graph K_2, every connected graph with bridges contain cut vertices as well.

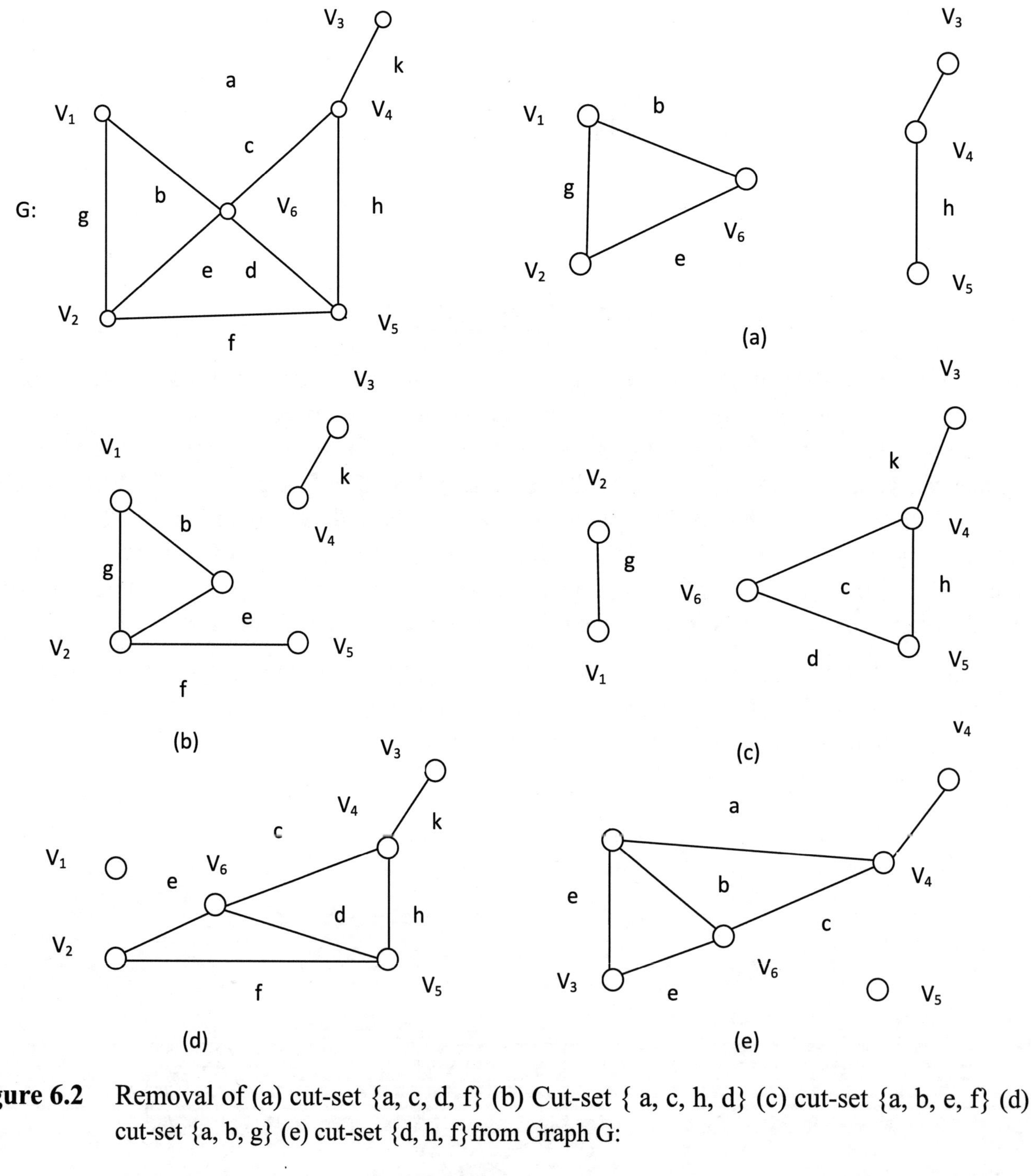

Figure 6.2 Removal of (a) cut-set {a, c, d, f} (b) Cut-set { a, c, h, d} (c) cut-set {a, b, e, f} (d) cut-set {a, b, g} (e) cut-set {d, h, f}from Graph G:

For a connected graph, there may be more than one cut-set. For instance, in Figure 6.2. the set of edges {a, c, d, f} is a cut-set. There are many cut-sets, such as {a, b, g}, {a, b, e, f}, and {d, h, f}. Edge{k} alone is also a cut-set, because one of itsproper subsets, {a, c, h} is also a cut-set.

A cut-set always "cuts" a graph into two. Therefore, a cut-set can also be defined as a set of edges in a connected graph whose removal reduces the rank of the graph by one. Another way of looking at a cut-set is this: if we partition all the vertices of a connected graph G into two mutually exclusive subsets, a cut-set is a minimal number of edges whose removal from G destroys all paths between these two sets of vertices. For example, in Figure 6.2 (a) cut-set {a, c, d, f} connects vertex set $\{v_{1,} v_2, v_6\}$ with $\{v_3, v_4, v_5\}$ i.e., A subset S of the edge-set of a connected graph G is called an edge cut-set or cut-set of G if

i) G-S is disconnected

ii) G-S_1 is connected for every proper subset S_1 of S.

REMARKS

1. If uv is an edge of an edge cut E', then all the edges having u and v as their ends also belong to E'.
2. No loop can belong to an edge cut.

There is a rich body of theorems concerning connectivity, many of which involves the number of disjoint paths joining a given pair of points in a graph. First we mention few theorems to illustrate the properties of cut-sets

THEOREM 6.1

An edge e of a graph G is a bridge iff e is not a part of any cycle in G .

Proof: Let e be a cut edge of G, andlet [X, X']={e} be the partition of V defined by G-e so that $x \in X$ and $y \in X'$. If e belongs to a cycle of G, then [X, X'] must contain at least one more edge, contradicting that{e} = [X, X']. Hence e cannot belong to a cycle.

Let us assume that e, a cut-edge of G be false. Then G-e is connected, and hence there exists an x-y path P in G-e. Then P U {e} is a cycle in G containing e.

Theorem 6.2

In a connected graph G, any minimal set of edges containing at least one branch of every spanning tree of G is a cut-set.

Proof: Let G be a connected graph and X' be the minimal set of edges containing at least one branch of every spanning tree of G. Consider the sub graph, G- X', after removing the edges in X'. Since the sub graph G-X' contains no spanning tree of G, G-X' is disconnected. Also, since X' is a minimal set of edges with this property, any edge e fromX' returned to G-X' will create at least one spanning tree. Thus the graph G- X'+e will be a connected graph. Therefore, X' is a minimal set of edges whose removal from G disconnects G. This, by definition is a cut set. Hence the proof.

THEOREM 6.3

A connected graph G is a tree if, and only if, every edge of G is a cut edge of G.

Proof: If G is a tree, there are no cycles in G. Hence no edge of G can belong to a cycle. By Theorem 6.1, each edge of G is a cut edge. Conversely, if every edge of a connected graph G is a cut edge of G, then G cannot contain a cycle, since the edges of a cycle are not cut edges of G. Hence G is a tree.

Cut Vertex is an analogue of a Bridge. A connected graph with cut vertices has the property that G can be disconnected by the removal of a single vertex. On the other hand, non separable graphs contain no cut vertices. However, a connected graph that is not complete has a vertex cut. In a sense, then non separable graphs are more highly connected than connected graphs with cut vetex, This suggests a way of measuring the connectedness of graphs. A subset v of the vertex set of G is called a vertex cut-set if (i) G-v is disconnected and (ii) G-v_1 is connected for every proper subset v_1 of v. For example, For the graph 6. 3, G, V = {v_1, v_3, v_5} is a vertex-cut set.

A vertex v of a graph is called a cut-vertex of G if ω (G-v) > ω (G) where ω (G) are number of components in G. i.e., A cut vertex breaks a graph into a sub graph having more connected components than G has,

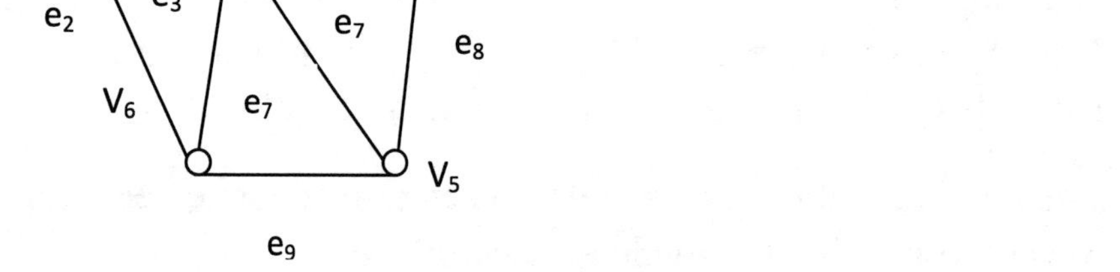

Figure 6.3

For example, for the graph in Figure 6.4, we have,

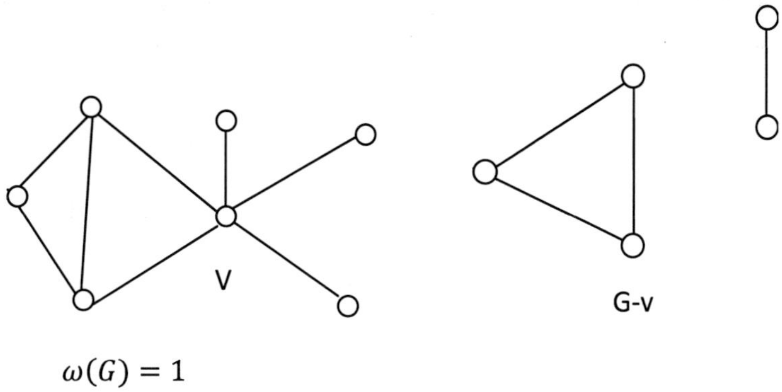

Figure 6.4 A graph after deleting cut vertex v

We now establish some facts about cut-vertices. Since a cut vertex in a graph G is a cut vertex of a component of G, we restrict ourselves to connected graphs.

THEOREM 6.4

The vertex v in a connected graph G is a cut-vertex if and only if there exist two vertices x and y in G such that every path between x and ypasses through v.

Proof : Suppose v is a cut-vertex of a connected graph G. Then G-v is disconnected, i.e., it has at least two disconnected components. Let x be any vertex in G-v and let y be any other vertex in G-v belonging to some component different from x's. Since G is connected, there exist some x-y path in G. However, there exist no x-y path. in G-v because x and y belong to different components of G-v. This implies that each x-y path in G passes through G, since G-v differs from G only in that G-v misses vertex v and all edges incident to v.

Conversely, suppose v is any vertex in a connected graph G and G contains some vertices x, y such that there is at least one x-y path in G and v lies on every x-y path in G. This assumption implies that any x-y path in G can no longer be a x-y path in G. Then the vertex v and all its incident edges are deleted from G. Thus, G-v is connected since it has vertices x and ythat are not connected. Hence, v is a cut-vertex of G.

THEOREM 6.5

Let v be a vertex incident with a bridge in a connected graph G. Then v is a cut vertex of G if and only if deg v ≥ 2.

Proof: Suppose that uv is a bridge of G. then deg v ≥ 1. Assume that deg v = 1. Since v is an end vertex of G, the graph G-v is connected and so v is not a cut vertex of G. Conversely, assume that deg v ≥ 2. Then there is a vertex w different from u that is adjacent to v. Assume, to the contrary, that v is not a cut vertex. thus G-v is connected, and so there is a u-w path P in G-V. However then, P together with v and the two edges uv and vw form a cycle containing the bridge uv.

This contradicts the theorem that an edge e in a graph g is a bridge if and only if e lies in no cycle of G. An immediate consequence of the above theorem is stated next.

Corollary 6.6

Let G be a connected graph of order 3 or more. If G contains a bridge, then G contains a cut vertex.

Proof: If v is a cut vertex in a connected graph G, then, of course, G-v has two or more components. If u and w are vertices in distinct components of G-v, then u and w are not connected in G-v. On the other hand, u and w are necessarily connected in G.

While the concepts of bridges and cut sets are parallel concepts, and have a number of similarities, they have some major difference as well. We have seen that it is possible for every edge of a connected graph G to be a bridge. Of course, G must then be a tree. A similar result is not true for cut vertices(i.e., every nontrivial tree connected graph must contain vertices that are not cut vertices). Our next result shows that not every vertex of a connected graph (with at least two vertices) is a cut vertex of G.

THEOREM 6.7

A vertex v of a connected graph G is a cut vertex of G if and only if there exist vertices u and w distinct from v such that v lies on every u-w path of G.

Proof: Suppose that v is a cut vertex of G. Then G-v is disconnectred. Let u and w be vertices in different components of G-v. It then follows by theorem above that every u-v path in G contains v. On the other hand, if G contains two vertices u and w such that every u-w path in G contains v, then there is no u-w path in G-v. Thus u and w are not connected in G-v, and so G-v is disconnected. Therefore, v is a cut vertex of G.

Corollary 6.8

Every nontrivial connected graph contains at least two vertices that are not cut vertices

Proof: Let u and v be vertices of a nontrivial connected graph G such that d(u, v) = diam(G). Since each of u and v is farthest from the other, it follows by above theorem that both u and v are not cut vertices of G.

6.2 EDGE CONNECTIVITY AND VERTEX CONNECTIVITY

Edge connectivity: The edge connectivity $\lambda(G)$ of a graph is the minimum cardinality of a set S of edges of G such that G-S is disconnected. i.e., the edge connectivity of a connected graph is the number of edges in a minimum cut-set (i.e., cut set with minimum number of edges) in the graph. The edge connectivity of a tree, for instance, is one. The edge connectivity of the graph of Figure 6.1 is one (since {k} is minimum cut set of the graph). For every graph of order n, $0 \leq \lambda \leq n-1$.

Note that $\lambda(G)$ is 0 if and only if G is disconnected or G is trivial, while $\lambda(G) = 1$ if and only if G is connected and contains a bridge. Therefore, every 1-edge connected graph is non trivial and connected and every 2-edge connected graph is a connected graph of order 3 or more that contains no bridges. i.e., For a nonnegative integer k, a graph G is said to be k-edge connected if $\lambda(G) \geq k$. In particular, the graphs $G_{3,}$ $G_{4,}$ and G_5 of Figure 6.1 are 2-edge connected graphs that are not 3-edge connected.

As noted earlier, a edge –cut in a graph G is the set U of edges of G that leaves G-U disconnected. An edge-cutof a connected graph G is minimal if no proper subset of X is an edge cut of G. If X is a minimal edge-cut of a connected graph G, then G-X contains exactly two components G_1 and G_2. Necessarily, then, X consists of all those edges of G joining G_1 and G_2. A edge-cut of minimum cardinality in G is called a **minimum edge –cut**. While every minimum edge- cut is a minimal edge -cut, the converse is not true.

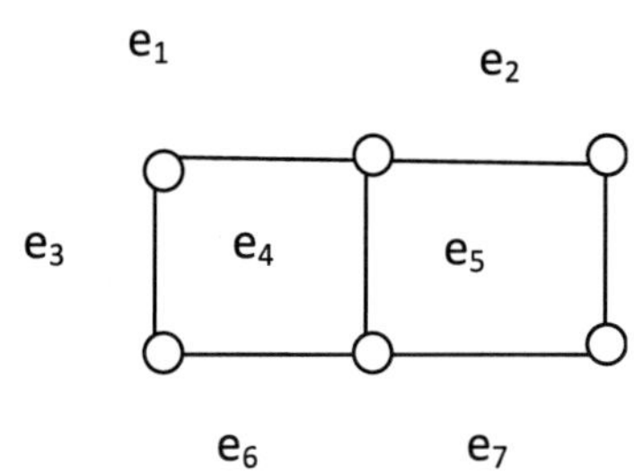

Figure 6.5 Graph illustrating edge-cuts

For the graph of Figure 6.5, though the sets $X_1=\{e_3, e_4, e_5\}$ and $X_3 = \{e_1, e_6\}$. are minimal edge cuts, the minimum edge cut is $X_3 = \{e_1, e_6\}$. The set $X_2=\{e_1, e_2, e_6\}$ is not a minimal edge cut as $X_3 = \{e_1, e_6\}$ is a proper subset of X_2.

Vertex connectivity: For a graph that is not complete, the *vertex connectivity* $\kappa(G)$ of a graph G is the minimum number of vertices whose deletion makes G a disconnected. Graph. i.e., the number of vertices in a minimum *vertex cut* is called the connectivity of the graph again, the vertex connectivity of a tree is one $\kappa(G)$. A vertex-cut of minimum cardinality in G is called a minimum vertex-cut. If U is a minimum vertex-cut in a non complete connected graph G, then G-U is disconnected and contain components $G_{1,}$ $G_{2, \ldots,}$ G_k, where $k \geq 2$. Every vertex $u \in U$ is adjaceant to at least one vertex in G_i for each I ($1 \leq i \leq k$); for otherwise U- {u} is also a vertex-cut which is impossible. When a connected graph G (on n vertices) does not contain K_n as a spanning sub graph, k is the connectivity of G if there exists a set of k vertices of G whose deletion results in a disconnected sub graph of G while no set of k-1 vertices has this property. For the graph of Figure 6.1, $\kappa(G_1) = \kappa(G_2) = \kappa(G_3) = 1$.

As noted earlier, complete graph does not contain vertex cuts. But for a graph that is not complete, the cardinality of a minimum vertex cut in G can never exceed the cardinality of any edge cut in G. Indeed the following theorem provides us with inequalities concerning vertex connectivity, edge connectivity, and minimum degree of graphs.

THEOREM 6.9

For every graph G, $\kappa(G) \leq \lambda(G) \leq \delta(G)$.

Proof: If G is disconnected or trivial, then, $\kappa(G) = \lambda(G) = 0$ and the inequalities hold; while if $G \cong K_n$ for some integer $n \geq 2$, then, $\kappa(G) = \lambda(G) = \delta(G) = n - 1$. Thus we may assume that G is a connected graph of order $n \geq 3$ that is not complete. Hence $(G)) \leq n - 2$. First we show that $\lambda(G) \leq \delta(G)$. Let v be a vertex of G with deg $v = \delta(G)$. Since the st of the $\delta(G)$ edges incident with v in G is an edge –cut of G, it follows that $\lambda(G) \leq \delta(G) \leq n - 2$.

It remains to show that $\kappa(G) \leq \lambda(G)$. Let X be a minimum edge-cut of G. Then $|X| = \lambda(G) \leq n - 2$. Necessarily, G-X contains exactly two components $G_1$and G_2. Suppose that the order of G_1 is v. Thus the order of G_2 is n - k, where $k \geq 1$ and $n - k \geq 1$. Consequently, every edge in X joins a vertex of G_1 and a vertex of G_2. We consider two cases.

Case 1: Every vertex of G_1 is adjacent to every vertex of G_2. Thus $|X| = k (n - k)$. Since $(k - 1)(n - k - 1) \geq 0$, it follows that $(k - 1)(n - k - 1) = k (n - k) - n + 1 \geq 0$. And so $\lambda (G) = |X| = k (n - k) \geq n - 1$. However, $\lambda (G) \leq n - 2$; so this case cannot occur.

Case 2: There exist vertices u in $G_1$and v in G_2 such that u and v are not adjacent in G. We now define a set U of vertices of G. For each $e \in X$, we select a vertex for U in the following way. If u is incident with e, then choose the other vertex in G_2 that is incident with e as an element of U; otherwise, select the vertex that is incident with e and belongs to G_1 as an element of U. Then $|U| \leq |X|$. Since $u, v \in U$and there is no u-v path in G-U, it follows that G-U is disconnected, and so U is a vertex-cut. Hence, $\kappa(G) \leq |U| \leq |X| = \lambda(G)$, as desired.

THEOREM 6.10

The edge connectivity of a graph G cannot exceed the degree of the Vertex with the smallest degree in G.

Proof: Let vertex v_i be the vertex with the smallest degree in v. Let $d(v_i)$ bethe degree of v_i. Vertex v_i can be separated from G by removing $d(v_i)$ edges incident on vertex v_i. Hence the theorem.

THEOREM 6.11

The vertex connectivity of any graph can never exceed the edge connectivity of G.

Proof : Let α denote the edge connectivity of H. Therefore, there Exists a cut set S in G with edges. Let S partition the vertices of G into subsets V_1 and V_2 By removing at most α vertices from V_1(or V_2) on which the edges in S are incident, we can effect the removal of S (together with the edges incident on the vertices) from G. Hence the theorem.

6.3 CYCLICAL EDGE CONNECTIVITY OF A GRAPH

In this section we introduce the parameter, cyclical edge connectivity of graphs. Unlike connectivity and edge connectivity, cyclical connectivity is not defined for graphs.

We define cyclical edge connectivity of G as the minimum number of edges of G whose deletion results in a graph having two components, each containing a cycle. It is denoted by $\lambda_c(G)$. It is clear that $\lambda \leq \lambda_c$. If a connected graph G is a union of edge disjoint cycles, then $\lambda_c(G)$ is even. This is because any edge cut has an even number of edges in common with a cycle. For example, the cyclical edge connectivity of Petersen graph is, 5.

6.4 SEPARABLE GRAPHS AND BLOCKS

In this section we turn our attention from connected graphs containing cut vertices to connected graphs without cut vertices. A nontrivial connected graph with no cut-vertices is called a non separable graph. Hence allof the graphs G_4, G_5, G_6, and $G_7$of Figure 6.1 are non separable. In addition K_2 is a non separable graph; indeed K_2 is the non separable graph of order 2 .

Separable graph: A connected graph is said to be separable if its vertex connectivity is one. That is, a connected graph is said to be separable, if there exists a sub graph g in G such that the complement of G and G have only one vertex in common. In a separable graph a vertex whose removal disconnects the graph is called a cut-vertex. In a tree, every vertex with degree greater than one is a cut vertex. Since non separable graphs of order 3 or more contain no cut- vertices, they contain no bridges; that is, every edge lies on a cycle.

Block: A ***block*** of a graph G is a maximal non separable sub graph of G. If G has no cut vertex, G itself is a block. Each block of G is an induced sub graph of G. Therefore, if G itself is non separable, then G has only one block namely, the graph G. On the other hand, if G is connected and has cut-vertices, then G has two or more blocks.

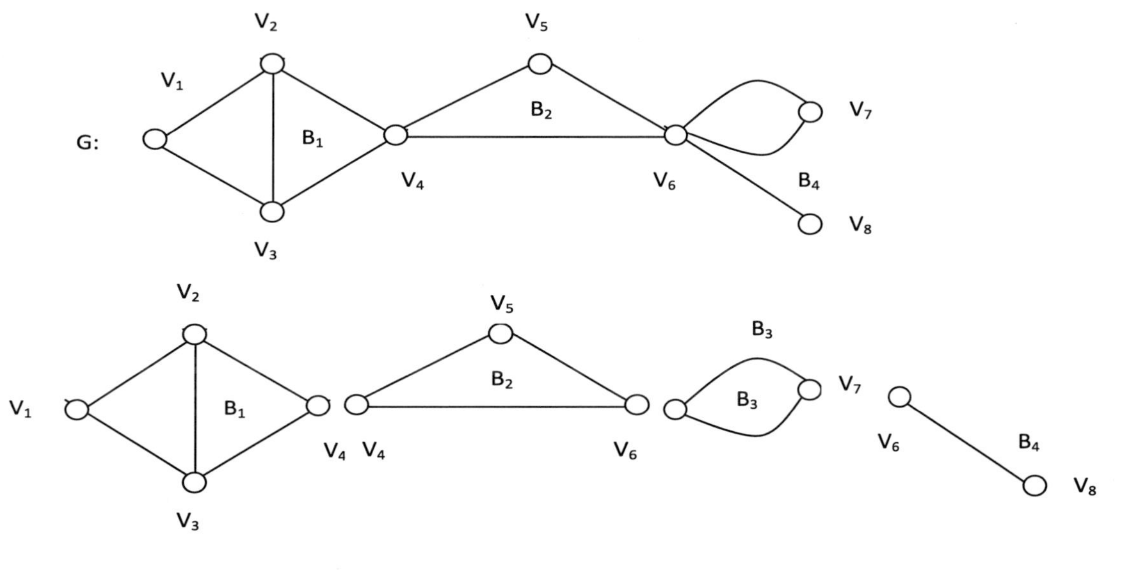

Figure 6.6 A connected graph and its Blocks

Figure 6.6 shows a connected graph G with two cut vertices v_4 and v_6 and four blocks B_1, B_2, B_3 and B_4. So just as a disconnected graph has special connected sub graphs called components, a connected graph with cut vertices contains special non separable sub graphs called blocks. The following theorem provides properties of blocks in a nontrivial connected graph.

SUMMARY

1. A **cut** in a graph G, is a partition of the vertices ofa graph, that splits it into two disjoint subsets. That is, vertex cut renders G disconnected. The **cut set** is the set of edges whose end points are in different subsets of the partition. In an unweighted graph, the size or weight of a cut denotes the number of edges crossing the cut. In a weighted graph, it implies the sum of weights of the edges crossing the cut. A cut is **minimum,** if the size of the cut is not largerthanthe size of any other cut. A cutis **maximum**, if the size of the cut is not smaller than thesizeof any other cut.

2. The **vertex connectivity** k(G) is the size of a minimal vertex cut. A graph is called **k-vertex connected** if its vertex connectivity is k or greater. The edge connectivity (G) is the size of a smallest edge cut and a graph G is called **k-edge connected** if its edge connectivity is k or greater.

3. A nontrivial connected graph with no cut-vertices is called a **non separable graph.** A **block** of a graph G, is a maximal non separable sub graph of G

EXCERCISES

1. For the following graph, . Find all the cut-sets of the graph?

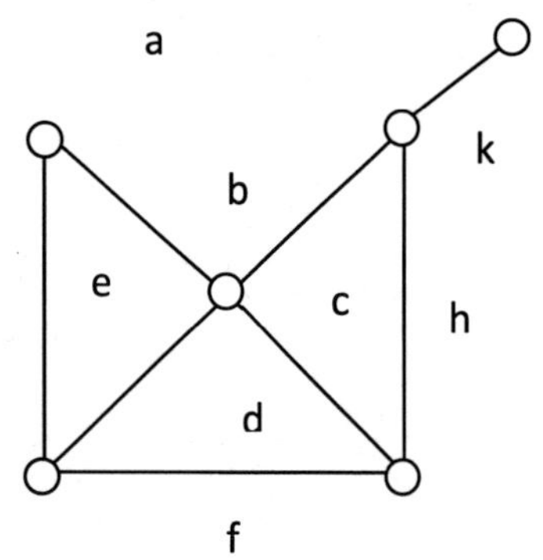

2. Find the edge connectivity of the following graph?

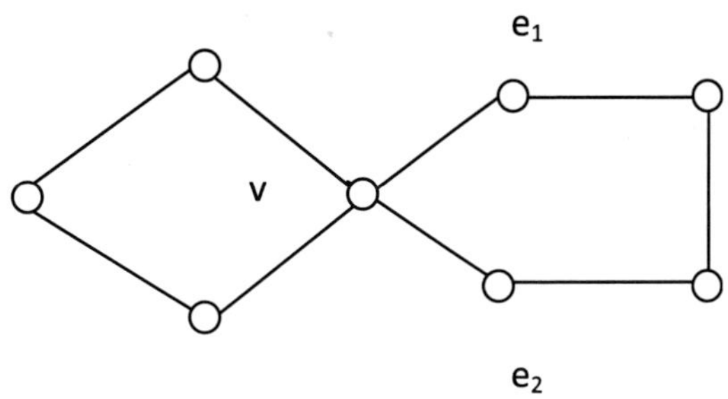

3. Find the minimum degree and edge connectivity of the graph shown below

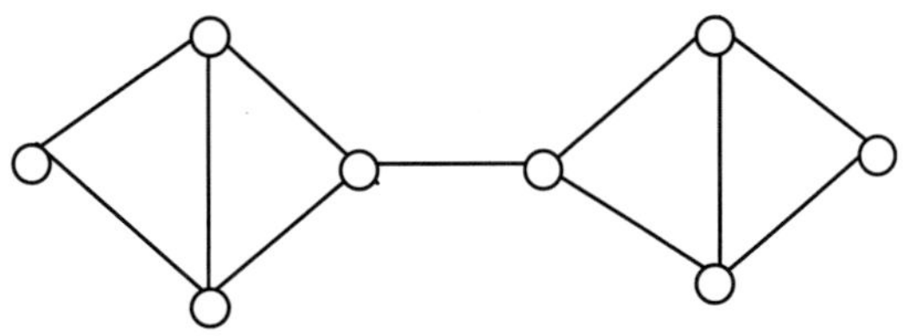

4. Find the edge and vertex connectivity of the following graph?

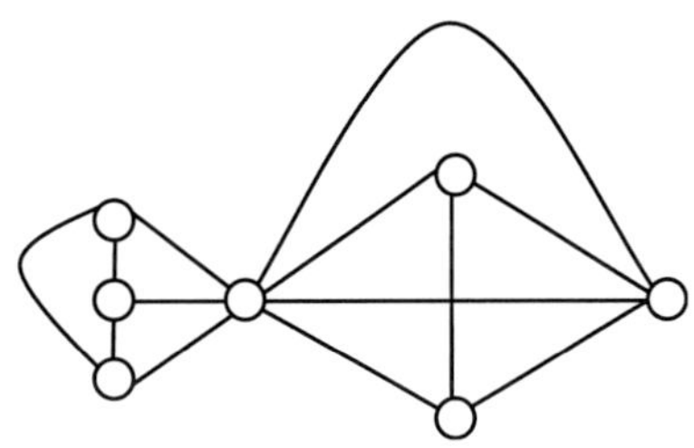

7

PLANAR GRAPHS

7.0 INTRODUCTION

The study of planar and non planar graphs, have contributed a great deal to the growth of graph theory. In addition, planarity and other related concepts are useful in many practical situations like VLSI circuit design, design of High speed highways, sub ways and Railroads, Irrigation canals design etc. For instance, in the design of electronic circuits, a good circuit design needs to be, on surface with fewer layers. That is, the solution requires the knowledge of whether or not the corresponding graph can be drawn in a plane. If not, drawing the same with the fewest possible crossings. Actually, the contributions by Euler, Kuratowski, and Whitney had been instrumental to the development of algebraic, topological, and computational techniques in graph theory. We present here some basic results on planar graphs, including the important characterization, the Kuratowski Theorem for planar graphs.

7.1 PLANAR AND NON PLANAR GRAPHS

In this section, we will study the question whether a graph can be drawn in the plane without edge crossings. There are always many ways to represent a graphs, . When is it possible to find the least one way to represent this graph in a plane without edge crossing? **A graph G is planar if there exists a drawing of G in which no edges intersectin a point other than a vertex of G, where each edge is a simple arc or Jordan arc** (A Jordan arc in the plane is a continuous intersecting curve whose origin and terminus coincide). Such a drawing of a planar graph G is called a plane representation of G. In this case we also say that G has been embedded in the plane. A drawing of a geometric representation of a graph on any surface such that no edges intersect is called embedding. . A plane graph is a planar graph already been embedded in the plane. That is a, planar graph is a plane graph if it can be drawn in the plane without edge crossings. Otherwise it is non planar.. Note that all trees are planar as also cycles and wheels. The Petersen graph is non planar.

For instance, consider the graph represented by Figure 7.1. The geometric representation shown in Figure 7.1, clearly is not embedded in a plane, because the edges e and f are crossing. But if we redraw edge f outside the quadrilateral, leaving the other edges, unchanged, we have, embedded the new geometric graph in the plane, thus showing that the graph whichis being represented by Figure 7.1 is planar. That is, a graph may be planar even if it is usually drawn with crossings, because it may be possible to draw it in a different way without crossings. On the other hand, we will not be able to draw any of the three configurations in Figure 7.2 on a plane without edges intersecting. The reason is that the graph with these different diagrams in Figure 7.2 represent a graph that is non planar.

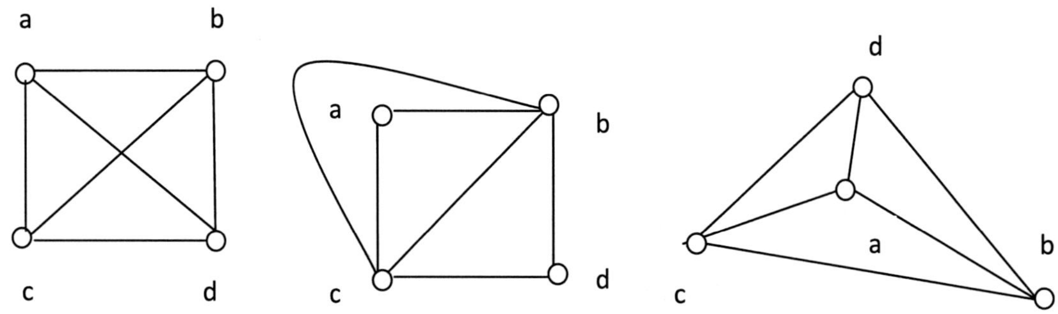

Figure 7.1 Planar graph K_4. Two plane representations of K_4

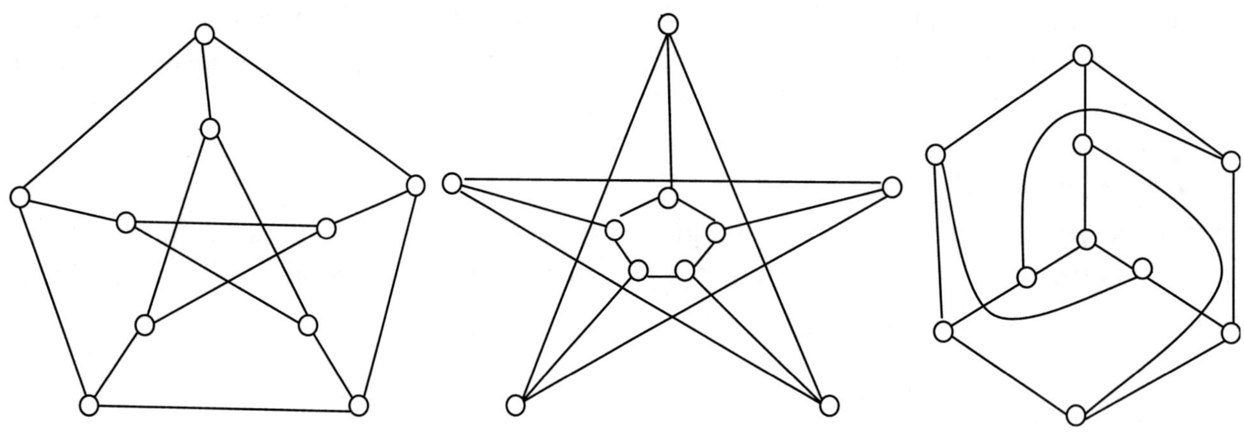

Figure 7.2 An example of Non planar graph(three isomorphic graphs)

Note 1: To show that a graph G is non planar we have to prove that of all possible geometric representations of G, none can be embedded in a plane.

A plane graph (planar representation)splits the plane into regions (also called faces, windows, or meshes), including an unbounded region. A region is characterized by the cycle (set of edges or vertices) that forms the boundary. These regions are connected portions of the plane. The area of a region may be finite or infinite. However, a planar graph has only one infinite region. Note that a region is not defined in a non planar graph or even in a plane graph not embedded in a plane. For example, the geometric graph in figure 7. 1 does not have regions. Thus, a region isa property of specific plane representation of a graph.

The portion of the plane lying outside a graph embedded in a plane is infinite in its extent. Such a region is called exterior region (also called infinite, unbounded, or outer region) for that particular plane representation. For instance the planar representation of the graph shown in Figure 7.3 splits into six regions. These are labeled in the Figure. Here r_6 is the exterior region. The subgraph of a plane graph with vertices and edges incident with a given region r make up the boundary of the region r. For example, a-e-d-a is the boundary of the region r_1. The degree of the region 'r' is the length of its boundary (or the number of edges incident with a region), denoted by deg (r). In counting the degree of a region a cut edge is counted twice. Thus, each edge of a plane graph G contributes two to the degree of the regionsor faces.

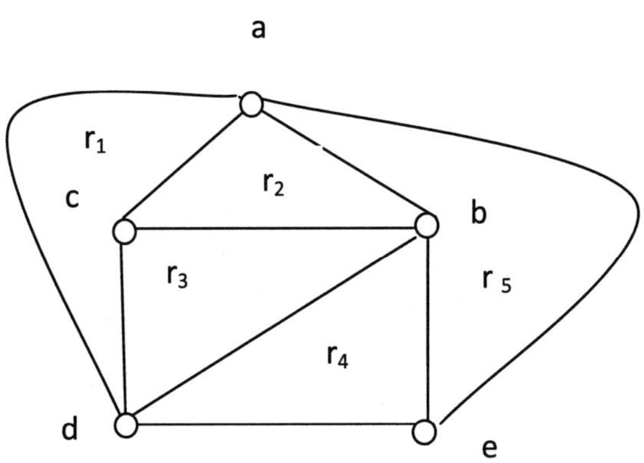

Figure 7.3 The regions of the planar representation of a Graph

In Figure 7.3, deg (r_4) = 3, deg (r_6) = 3, . etc. Like other regions, the infinite region is also characterized by a set of edges (or vertices). Clearly, by changing the embedding of a given planar graph, we can change the infinite region. For instance, Figure 7.4 are two different embeddings of the same graph. The finite region v_1, v_3, v_5 in Figure 7.4 (a) becomesthe infinite region in Figure 7.4(b). Another observation is useful: If G is a connected plane graph with at least three edges, then the boundary of every region of G has at least three edges. Also, A connected graphis a tree, if and only if, it has only one region. Or conversely, if a connected plane graph has only one region, then it must be a tree.

Remarks

1. Fora connected plane graph G, each edge of G belongs to one or two regions of G. A cut edge on the other hand belongs to exactly one region. That is, an edge that is not a cut edge belongs to exactly two faces.

2. Fora connected plane graph G, the boundary of each region is a closed walk in which cut edge is traversed twice. When there are no cut edges, the boundary of each region is of G is a closed trail in G. However, if G is a disconnected plane graph, then the edges and the vertices incident with the exterior region will not define a trail.

3. In testing the planarity of a graph G, one may delete multiple edges and loops of G, if any. This is so because, if a graph H is nonplanar, the removal of loops and parallel edges of H results in a sub graph of H, which is also non planar.

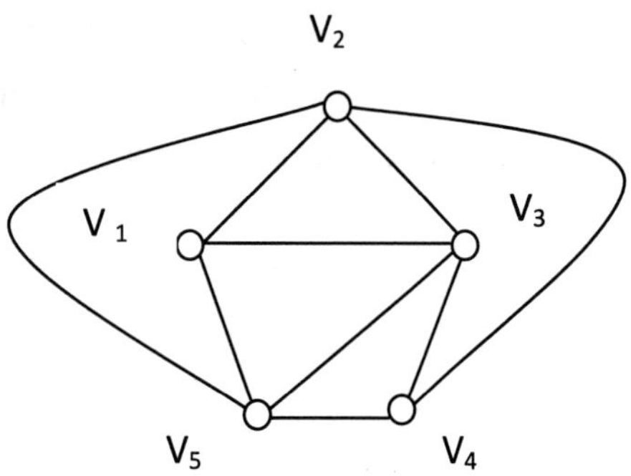

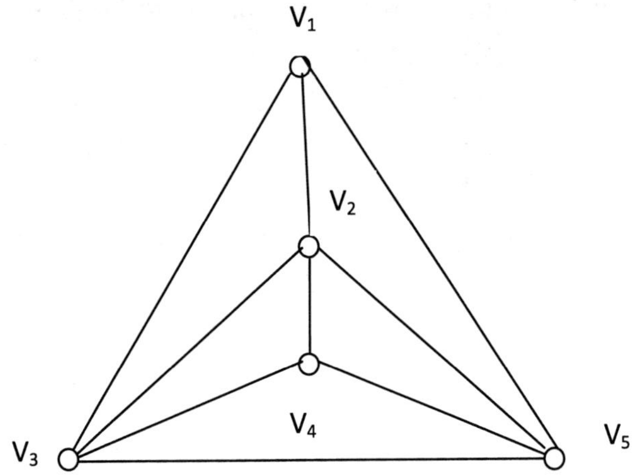

Two different embeddings

Figure 7.4(a) **Figure 7.4(b)**

We have noted that, a planar graph may have more than one plane representation. That is., a planar representation of a graph splits the plane into regions, including an unbounded region. A natural question that would arise is whether the number of regions (faces) is the same in each representation? The answer to this question is provided by Euler. Eulerin 1752, by his remarkable formulae called Euler's formulae, showed that all planar representation of a graph splits the plane in to the same number of regions. He accomplished this by finding a relationship between the number of regions, number of vertices and the number of edges of a planar graph. That is., If G is a connected planar graph then any drawing of G in the plane as a plane graph will always form $r = m - n + 2$ regions, including the exterior region, where n, m, r denote the number of vertices, edges, and regions respectively. Indeed, Theorem 7.1, known as Euler's formulae, gives the number of regions in any planar graph.

THEOREM 7.1

If G is a connected plane graph of order n, size m, and having r regions, then $n\text{-}m + r = 2$.

Proof: (By inductionon r, the number of regions). If $r = 1$, then G is a tree and $m = n - 1$. Hence, $n - m + r = 2$. let us assume that the result is true for all connected planar graphs G with $r + 1$ regions and n vertices. Deletion of an edge common to both regions of G results in the formation ofa graph G^1 with the same number of vertices. The resulting graph G^1 has one fewer edge and one fewer region, since two previous regions have been combined by removal of an edge. Also, since e is not a cut edge of G, G-e is connected. Further, the number of regions of G-e is r-1. So, by the induction hypothesis, $n-(m-1) + (r-1) = 2$. This completes the proof of the theorem.

Example 1:

Suppose that a connected planar simple graph has 20 vertices, each of degree 3. Into how many regions does a representation of this planar graph splits the plane?

Solution

This graph has 20 vertices, each of degree 3. So n = 20. Because the sum of the degrees of the vertices, 3n = 3. 20 = 60, is equal to twice the number of edges, 2m, we have 2m = 60, or m = 30. Comsequently, from Euler's formulae, the number of regions is, r = m - n + 2 = 30 - 20 + 2 = 12.

Euler's formulae and a method of proof by contradiction will solve the following problem.

Example 2:

Prove that there does not exist a polyhedral graph with exactly seven edges.

Solution

If there were such a polyhedral graph with m =7, then $3r \leq 2m = 14$. Since each region has degree ≥ 3. Moreover, each vertex has degree ≥ 3, so that $3n \leq 2e = 14$. Thus, $r \leq 4$ and n = m + 2 = 9, a contradiction. Therefore, we conclude that a polyhedral graph with 7 edges does not exist.

The Euler identity can be used to establish some inequalities that must be satisfied by planar graphs and has many usefuland interesting consequences. One of these tells us that planar graphs cannot have too many edges.

Corollary 7.2

If G is a planar graph with at least three vertices, then $m \leq 3n-6$. The proof of corollary 7.2 is based on the concept of the degree of a region, which is defined to be the number of edges on the boundary of this region. When an edge occurs twice on the boundary, it contributes two to the degree. The proof corollary 7.2 can now be given.

Proof: A connected simple planar graph drawn in the plane divides the plane into regions, say r, of them. The degree of each region is at least three. In particular, note that the degree of the unbounded region is at least three because there are at least three vertices in the graph. Note that the sum of degrees of the regions is exactly twice the number of edges in the graph, because each edge occurs on the boundary of a region exactly twice(either in two different regions, or twice in the same region). Because each region has degree greater than or equal to three, it follows that

$$2m=\sum_{all\ regions\ R} \deg(R)\geq 3r$$

Hence, $(2/3)\ m \geq r$. Using r= m –n + 2 (Euler's Formulae), we obtain $m - n + 2 \leq (2/3)\ m$. It follows that $m/3 \leq n-2$, This shows that $m \leq 3n-6$. The above resultis not valid if n =1, Or 2. Also, the condition that $m \leq 3n-6$ is not sufficient for planarity of a simple connected graph.

For example, the Petersen graph has m=15, n=10 and hence m ≤ 3n-6. But the graph is not planar. However, This corollary can be used to show that K_5 is non planar .

Example 3:

Show that K_5 is non planar.

Solution

The graph K_5 has order 5 and size 10. However the in equality m ≤ 3n–6. Is not satisfied for this graph because m =10 and 3n–6 = 9. Therefore K_5 is non planar.

Example 4:

Is $K_{3, 3}$ is planar?

Solution

Assume, that $K_{3, 3}$ is a planar graph. Since n = 6 and m = 9, it follows by the Euler identity that n – m + r = 6 – 9 + r = 2 and so r = 5. Let R_1, R_2, …. , R_5 be the five regions, and let m_i be the number of edges on the boundary of R_i (1 ≤ I ≤ 5). Since $K_{3, 3}$ has no triangles, m_i ≥ 4 for 1 ≤ I ≤ 5 and because $K_{3, 3}$ contain no bridges it follows that $2m = \sum_{i=m}^{5} m_I \geq 20$. And so m ≥ 10. This is a contradiction.

Note 2:

For any edge e of K_5, K_{5-e} is planar. So also is the casewith $K_{3, 3}$. Observe that the graphs K_5 and $K_{3, 3}$ have some features in common. They are:

i. Both are regular graphs.

ii. Removal of a vertex or an edge from each graph results in a planar graph.

iii. Contraction of an edge results in a planar graph.

iv. K_5 is a non planar graph with the smallest number of vertices, whereas $K_{3,3}$ is a non planar graph with the smallest number of edges. Hence, any non planar graph must have at least 5 vertices and 9 edges.

THEOREM 7.3

For any simple planar graph δ(G) ≤ 5,

Proof: Suppose that G is a graph, every vertex of which has degree six or more. Let G have order n and size m. Certainly n ≥ 7. Then $2m = \sum_{v \in V(G)} \deg v \geq 6n$. Thus m ≥ 3n > 3n–6. This contradicts m ≤ 3n–6, and by corollary 7.2, G is non planar.

Corollary 7.4

If G is a loop free connected planar graph with n (≥ 3) vertices and m edges and number of circuits of length 3, then m

Proof: Since degree of each region in a planar graph is at least 4, the total number of edges as the part of boundary of all regions is at least 4r. In general, each edge of the planar graph is part of the boundary of two regions, and therefore, total number of edges around all the regions is 2e. Consequently, 2e ≥ 4r or e ≥ 2r. Substituting this value of r in the Euler's formulae, we have

$$2 = n\text{-}m + r \textbf{ or } r = 2\text{-}n + m.$$

$$2r = 4\text{-}2n + 2m \leq m \textbf{ or}$$

$$\mathbf{m \leq 2n\text{-}4.}$$

THEOREM 7.5

A complete graph K_n is planar iff n ≤ 4.

Proof: It is east to see that K_n **is planar** for n = 1, 2, 3, 4. Thus, we have only to show that K_n is non planar if n ≥ 5, and for this it suffices to show that K_5 is non planar r. We prove this by an indirect argument. If k_5 were planar, then r = 2–n+m = 2–5+10 = 7. But since K_n is simple and loop free, we would also have 3r ≤ 2e which in this case would imply that 3. 7 ≤ 2. 10 =20. , an obvious contradiction. One could also obtain asimilar contradiction by appealing to the in equality m ≤ 3n–6.

THEOREM 7.6

A complete bipartite graph is planar iff m ≤ 2 or n ≤ 2.

Proof: It is easy to see that $K_{m, n}$ is planar if m ≤ 2 or n ≤ 2. Now let m ≥3 or n ≥ 3. to prove that $K_{m, n}$ is non planar it suffices to prove that $K_{3, 3}$ is non planar. Since $K_{3, 3}$ has six vertices and nine edges, if $K_{3, 3}$ were planar, Euler's formulae would give that r = 2-n+m= 5. Since $K_{3, 3}$ is bipartite there can be no cycles of odd length. Hence each cycles has length ≥ 4 and thus the degree of each region would have to be greater than or equal to 4. But then we would have to have 4r ≤ 2m or 20= 4. 5 ≤ 2. 9 =18, a contradiction.

7.2 MAXIMAL PLANAR GRAPH

A graph G is **maximal planar** if G is planar but the addition of an edge between any two non adjacent vertices of G results in a nonplanar graph. Another way to say is that a graph G is **maximal planar** if G is planar but G is not a proper spanning subgraph of any other planar graph.

For $1 \leq n \leq 4$, the only maximal planar graph of order n is $K_{n..}$ That is, any planar graph is a spanning sub graph of a maximal planar graph. More over, if G bar is a plane embedding of a planar graph G with at least three vertices, and if e = uv is a cut edge of G bar embedded in a region of G barthen there exists a vertex w on the boundary of region r such that an edge uw or vw can be drawn in r so that either G bar + vw or G bar + uw is also a plane graph. Thus all of the graphsin Figure 5.5 are maximal planar. Necessarily, then, if a maximal planar graph G of order $n \geq 3$ and size m is drawn as a plane graph, then the boundary of every region of G is a triangle and m = 3n–6.

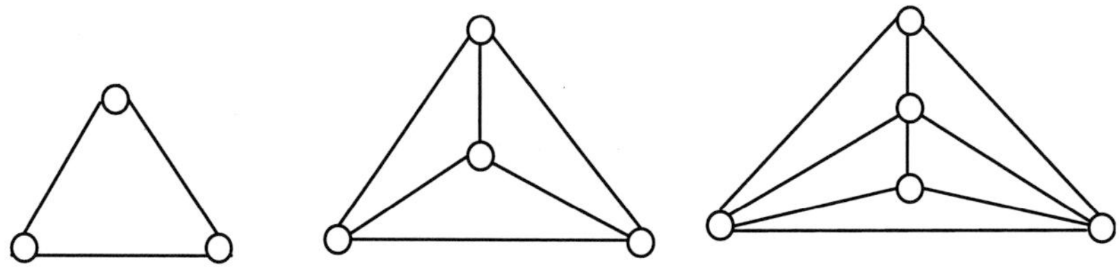

Figure 7.5 Maximal planar graphs.

7.3 DUAL OF A GRAPH

Given a plane graph G, one can form a new graph G^* in the following way. Corresponding to each region r of G, take a vertex r^* and an edge e^*corresponding to edge eof G. Then edge e*joins vertices r^* and g^* in H if, and only if, edge e is common to the boundaries of the regions r and g in G. Graph H is then called the dual of G. If e is a cut edge of G embedded in region r of G, then e^* is a loop at r^*... H is a planar graph, and there exists a natural way of embedding H in the plane. Vertex f^* corresponding to region f, is placed in region rof G . edge e^*, joining r^* and g^*, is drawn so that e^* crosses e once and only once and crosses no other edge. (Figure 5.6). This embedding is the canonical embedding of H denoted by G^*. The definition of dual implies that $m(G^*) = m(G)$, n (G^*) =r (G) and $d_G(r^*) = d_G$ (r) denotes the degrees of the region r of G. .

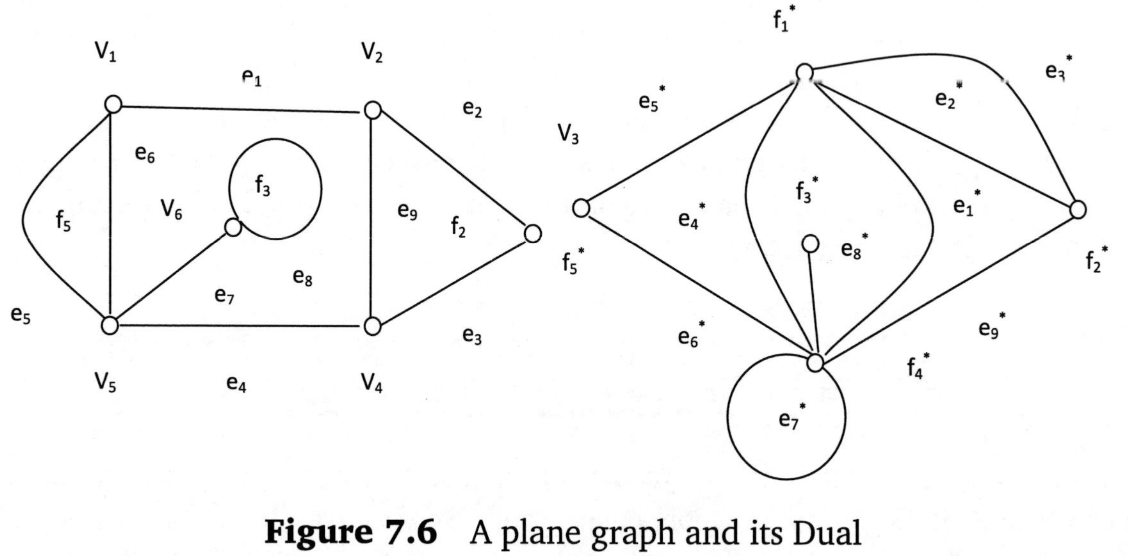

Figure 7.6 A plane graph and its Dual

Observations on (Geometric) Dual of a graph

1. An edge forming a self loop in G yields a pendent edge in G^*.
2. A pendent edge in G yields a self loop in G^*.
3. Edges that are series in G produces parallel edges in G^*.
4. Edges that are parallel in G produces series edges in G^*.
5. The dual of the dual graph is the graph itself. Thus G and G^* are duals to each other** is isomorphic toG, if and only if, G is connected. Graph isomorphism does not preserve duality. That is, isomorphic planeGraphs may have non isomorphic duals.
6. G^* is connected whether G is connected or not. For a given planar graph G, the graph G^* neednot beUnique.
7. G^* is planar.
8. $|E^*| = |E|$ and $|R^*| = |V|$.
9. $deg_G(r^*) = deg_G(r)$ for each region v of G where r^* v^* (G^*) and r r(G).
10. The nullity of $G(G^*)$ = The rank of $G^*(G)$.
11. A necessary condition and sufficient condition for two planar graphs G_1 and G_2 to be duals of each other is as follows: There is a one to one correspondence between edges in G_1 and theedges in G_2such that a set of edges in G_1 form a cyclethe corresponding set in G_2 formsa cut set.

Note: 1: A planar graph isomorphic to its own dual iscalled self dual graph.

Example 5:

$K_4$is self dual as was shown in Figure 7.7

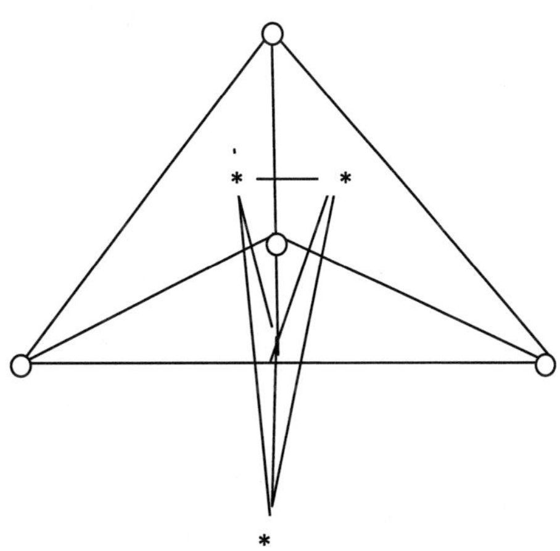

Figure 7.7 Self dual graph

7.4 HOMEOMORPHIC GRAPHS

Two graphs G_1 and G_2 are **Homeomorphic**, if G_1 and G_2 can be reduced to isomorphic graphs by performing a sequence of series reductions. That is, they are **Homeomorphic** if each can be obtained from the same graph by adding appropriate vertices (necessarily of degree 2) to edges. By this definition, we can say that any graph is Homeomorphicto itself. Also, graphs G_1 and G_2 are Homeomorphic if G_1 can be reduced to a graph isomorphic to G_2 or if G_2 can be reduced to a graph isomorphic to G_1 .

Example 6:

The graphs shown in Figure 7.8 are Homeomorphic since both are obtainable from the same graph G by adding vertices (necessarily of degree 2) to one of its edges.

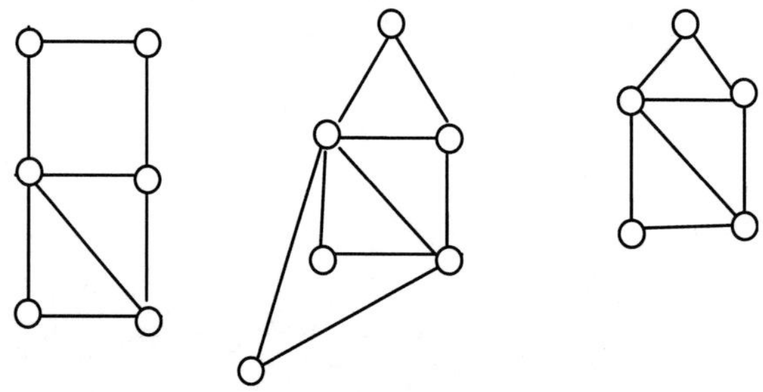

Figure 7.8 Two Homeomorphic graphs G_1 and G_2 of G

Example 7:

We show two Homeomorphic graphs (Figure 7.9), each obtained from K_5 by adding vertices to edges of K_5. In each case, vertices of K_5 are shown with solid dots.

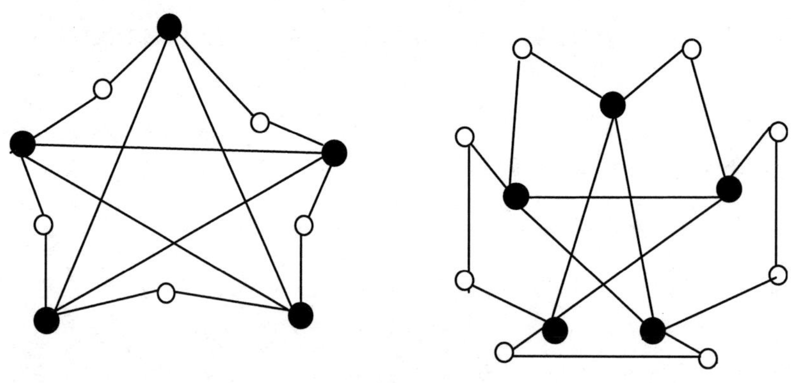

Figure 7. 9 Two Homeomorphic graphs obtained from K_5

THEOREM 7.6

A graph G has a dual if and only if it is planar

Proof: We have to show that no non planar graph has a dual. Let G be a non planar graph. Then G contains asub graph Homeomorphic to K_5 and $K_{3,3}$. But K_5 and $K_{3,3}$ have no dual. Moreover, series edges of a graph yields only parallel edges in the dual graph and hence the parallel edges in the Homeomorphic images of K_5 or $K_{3, 3}$ does not support to write the dual. Hence G has no dual.

THEOREM 7.7 (KURATOWSKI THEOREM):

A graph G is planar if, and only if, it has no sub graph Homeomorphic to K_5 or $K_{3,3}$.

Proof: Since K_5 or $K_{3, 3}$ are non planar (as was shown above), it follows that if a graph contains a sub graph Homeomorphic to either of these, it is also non planar. The proof of the converse, namely that every non planar graph contains a sub graph Homeomorphic to K_5 or $K_{3, 3}$. Is complicated and will not be given here. Examples below show how Kuratowski theorem is proved.

Example 8:

is Petersen Graph shown in Figure 5.10 (a) planar.

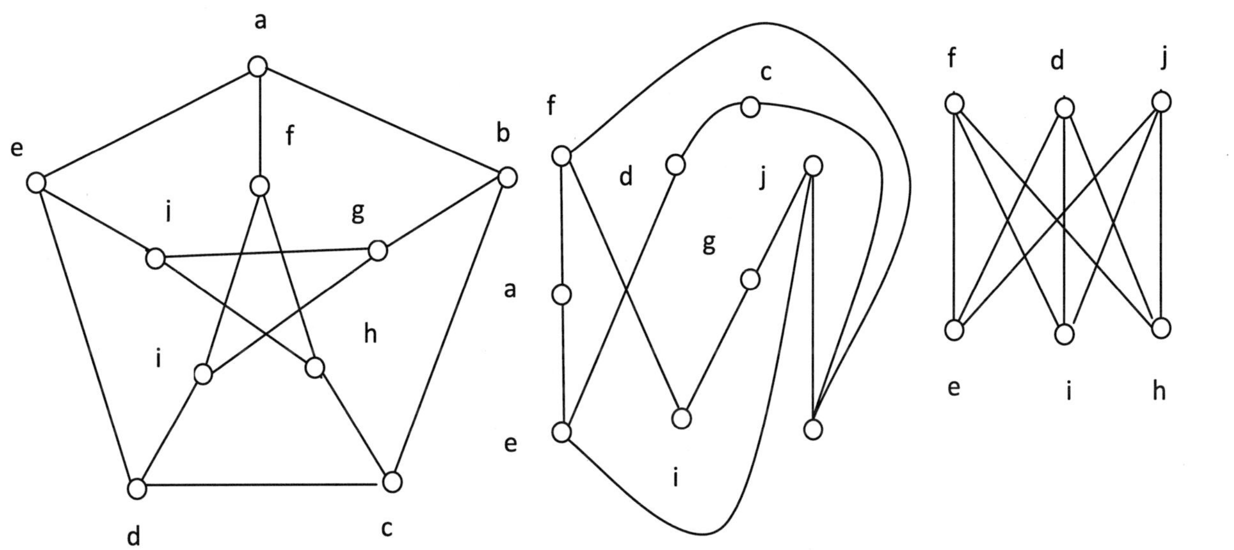

Figure 7.10(a) The Petersen graph **(b)** Sub graph H Homeomorphic to $K_{3,3}$ **(c)** and $K_{3, 3}$.

Solution

The sub graph H of the Petersen graph obtained by deleting b and the three edges that have b as an end point, shown in Figure (b), is Homeomorphic to $K_{3,3}$. With vertex sets {f, d, j} and {e, i, h}, because it can be obtained by a sequence of elementary sub divisions, deleting {d, h} and adding {c, h} and {c, d}, deleting {e, f} and adding {a, e} and {a, f}, and deleting {i, j} and adding {g, i} and {g, j}. Hence the Petersen graph is not planar.

SUMMARY

1. A **planar graph** is a graph drawn in the plane in such a way that any pair of edges meet only at their End vertices. A **planar graph** is a graph which is isomorphic to a plane graph. That is, it can be redrawn as a plane graph.

2. A plane graph divides the plane into connected parts called **regions**. In every plane graph, there is always one region that is unbounded, which is the **exterior region.** The sub graph, whose vertices and edges, are incident with a given region R, is the **boundary** of R. A connected plane graph with only one region represents a Tree.

3. For a connected plane graph with order n, size m, andregion r, **Euler's formulae**, $n\text{-}m + r = 2$, gives thenumber ofthe regions in the connected plane graph.

4. A **planar graph is maximal**, if it is not possible to add an edge such that the graph is still planar.

5. The dual of a plane graph G is a graph that has a vertex corresponding to each face of G, and an edge joining two neighboring faces for each edge in G.

6. By **Kuratowskii's theorem**, A graph G is planar if, and only if, it has no sub graph Homeomorphic to K_5 or $K_{3, 3}$.

EXCERCISES

1. Show that each of the graphs in figure below are planar by drawing it as a plane graph. VerifyEuler Identity for each Graph?

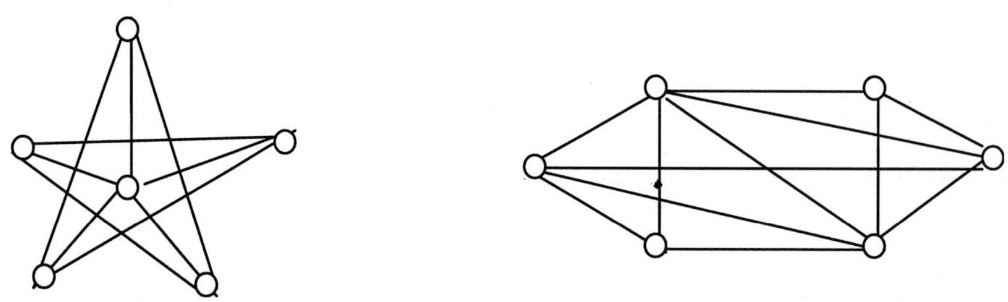

2. Detemine with explanation whether the graph in figure below is planar?

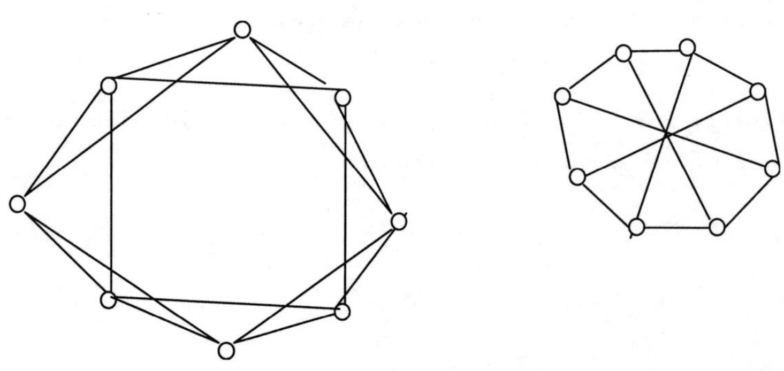

3. Check whether the graphs are Homeomorphic?

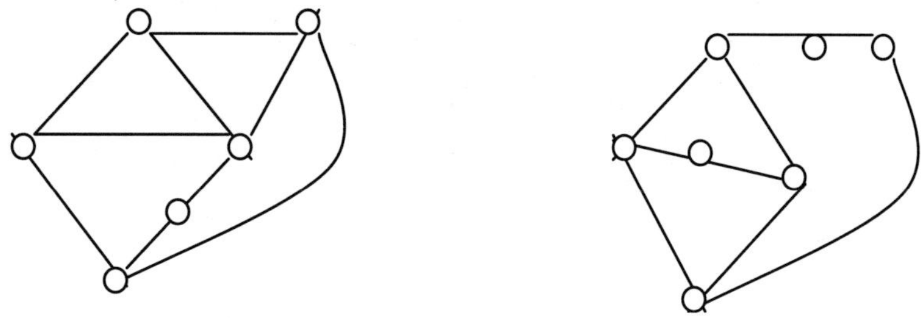

4. Show that the following graph is not planar by finding a sub graph Homeomorphic to either K_5 or K_{33}?

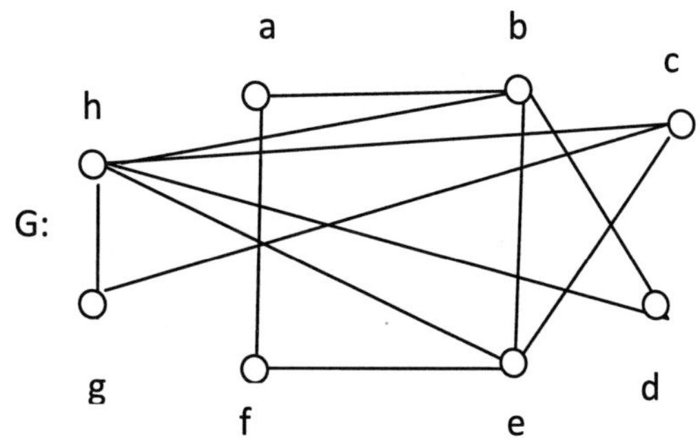

5. Show that the following graph is non planar?

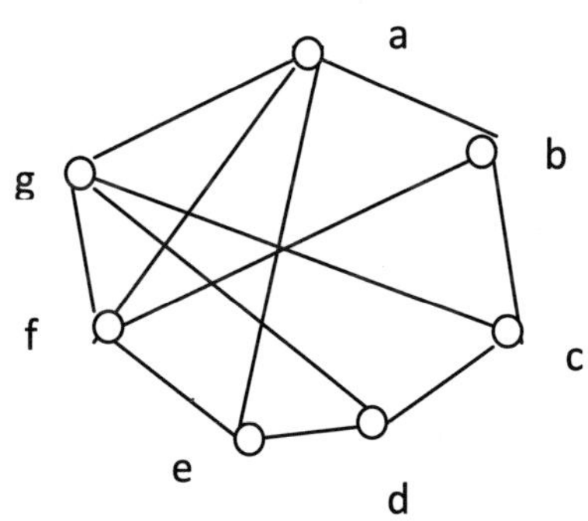

6 Show that the graphs shown below are Homeomorphic?

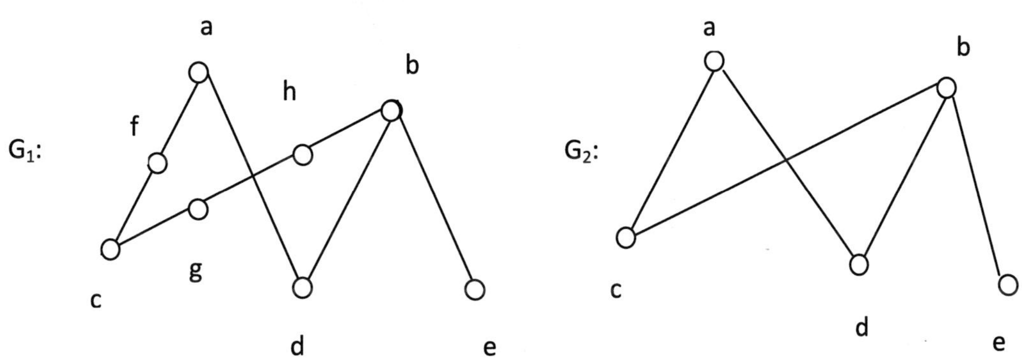

7. Draw the given planargraph without crossings?

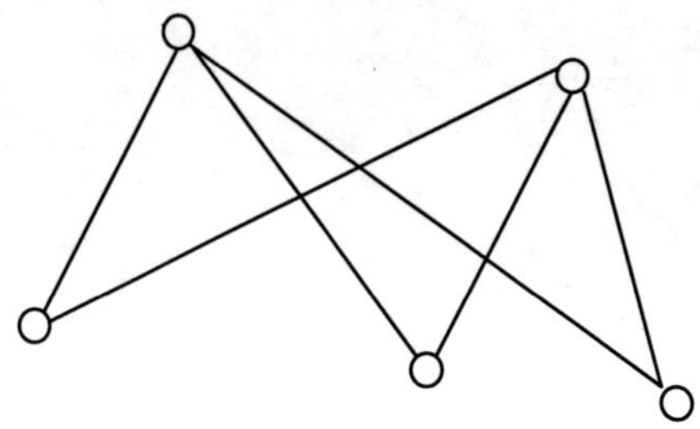

8. Determine whether the Graph shown in figure below hasa sub graphHomeomorphic to K_{33}. ?

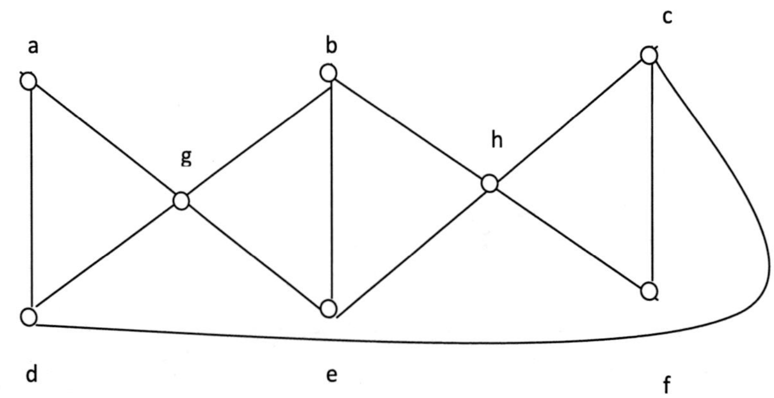

9. Use Kuratowskii's Theorem to determine whether the given graph is planar?

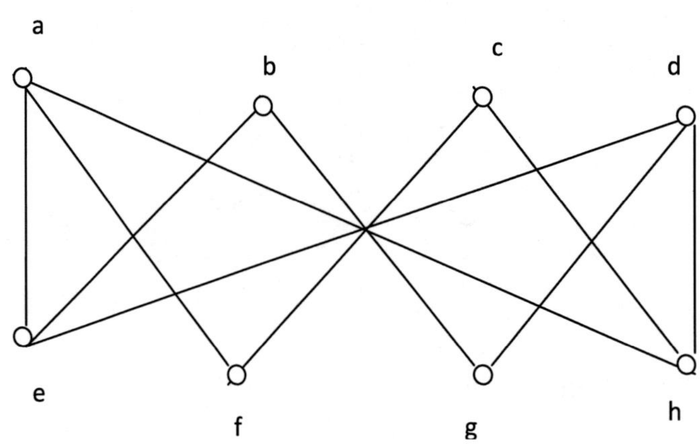

8

NETWORKS AND FLOWS

In this section we discuss networks model. The network may be a transportation network through which commodities (tangible like oil drums, logistics, automobiles, finished goods, or money or intangibles like, disease, information, or protein structure or any number of other possibilities) flow. Flow is a network and is made of directed graphs. In each case, the problem is to determine a maximal flow. Maximizing the flow in a network is a problem that belong s to graph theory and operations research.

8.1 NETWORK MODELS

A (transport) network is a simple, weighted, directed graph G (that is weakly connected and contains no loops) satisfying the following:

1. There are two distinguished vertices S and D of G, called the **source** and **sink** of G respectively i.e., in G, only one vertex with no incoming edge is called assource, S and only one vertex with no outgoing edge is called as sink, D (In general source has in degree '0' and sink has out degree '0'). The other vertices are called **intermediate vertices** (the flow in to an intermediate vertex **u** is equal to the flow out of **u**).
2. The weight C_{ij} of the directed edge (i, j) is a non negative number. i.e., every edge is associated with non negative real numbers called as **capacity** of that edge.
3. The underlying undirected graph is connected.

8.2 FLOW

8.2.1 DEFINITION

Let (G, C) be a network with source S and sink D. Assume that the capacity function C is defined on the edges of G. A flow in G is a non negativereal valued function F defined on the edges of G such that

(i) $0 \le F(e) \le C(e)$ for each edge $e \in E(G)$.

(ii) For each intermediate vertex, the sum of all values F(x, y) such that $y \in A(x)$

Must equal the sum of all values F(z, x) such that $z \in B(x)$, and

(iii) F(e) = 0 for any edge e incident to the source Sor incident from the sink D.

Condition (i) is called the **capacity constraint**, and it insures that the amount of flow along an edge does not exceed the capacity for that edge. Condition (ii) requires that the sum of all incoming flows at x is equal to the sum of all outgoing flows at x.

In symbols, condition (ii) becomes

$$\text{(ii)} \sum_{y \in A(x)} F(x,y) = \sum_{z \in B(x)} F(z,x) \qquad (6.1.1)$$

For every $x \in V(G) - \{S, D\}$

Thus, $\sum_{(y \in A(x))} F(x, y)$ is the flow out of x $\sum_{(y \in B(x))} F(x, y)$ and is the flow into x. Moreover, the net flow out of a vertex x is defined to be $\sum_{(y \in A(x))} F(x, y) - \sum_{(y \in B(x))} F(x, y)$, while the net flow into x, is of course, $\sum_{(y \in B(x))} F(x, y) - \sum_{(y \in A(x))} F(x, y)$. Thus, conditions (i) and (ii) assure that for an intermediate vertex x the **net flow out of x** equals the **net flow into x**, and this common value is zero. In other words no flow is created or destroyed at an intermediate vertex.

Condition (iii) insures that the flow moves from source S to sink D and not in reverse direction. Note that every transport network has at least one flow, namely, the zero flow, where F(e) = 0 for each edge $e \in (G)$. We label each edge e as (C(e), F(e)) where C(e) is the capacity of e and F(e) is the flow along e.

Note 1: A flow in a network assigns a flow in each directed edge which does not exceed the capacity of the edge.

THEOREM 8.2.1

Let S and D be the source and sink, respectively, of a network (G, C). Let F be a flow in G. Then the flow out of S = F(S(V(G)) = (F(V(G), D) = the flow into D.

Proof: We first observe that

$$F(V(G),V(G)) = \sum_{x \in V(G)} F(x, V(G)) \quad (6.1.2)$$

$$= \sum_{x \in V(G)} F(V(G), x)$$

By condition (ii) of definition 6.2.1, F(x, V(G)) = F(V(G), x) for each x ∈ V(G) \ {S, D}. Thus the above sum becomes

$$F(S, V(G))+F(D, V(G))=F(V(G), S)+F(V(G), D) \quad (6.1.3)$$

But in addition, we know that F(V(G), S) = 0 = F(D, V(G)) since there is no flow into S and no flow out ofD. Thus equation (6.1.3) becomes F(S, V(G)) = F(V(G), D). Hence the theorem is proved.

8.3 VALUE OF FLOW

DEFINITION 8.3.1

Let F be a flow definedon the transport network (G, C). Let $a_{1,}$ a_2, a_3, ... a_m denote all vertices that are incident from the source S, andlet b_1, b_2, ...b_n denote all vertices incident to the sink D. Thus, { $a_{1,}$ a_2, a_3, ... a_m } and B(D) = { b_1, b_2, ...b_n }. The quantity is called **value** of the flow F. Whenever the flow along an edge is equal to the capacity of that edge, then that edge is said to be **saturated.** Otherwise it is **unsaturated.**

$$|F| = F(S,a_1) + F(S,a_2)+.............+F(S, a_m)$$

$$= F(S,V(G)).$$

$$= F(b_1,D) +F(b_2,D)+...........+F(b_n, D)$$

$$= F(V(G),D).$$

Example 8.3.2:

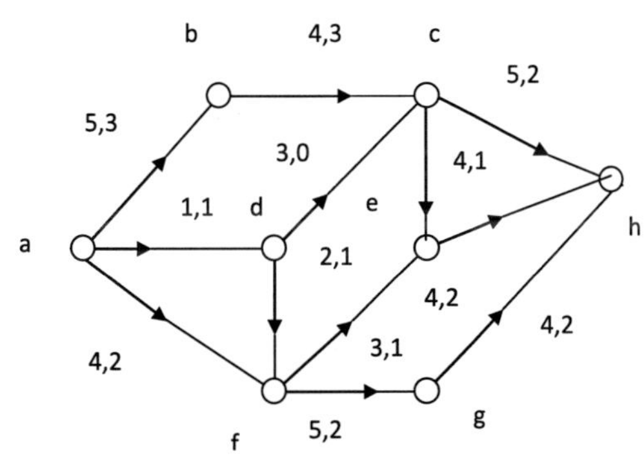

Figure 8.1

Here, the first quantity represents the capacity and the second quantity represents the flow. The flow out of the **source** a is F(a, b)+F(a, d)+F(a, f) = 3 +1+2 = 6 and the flow into the **sink** z is F(g, z) +F(e, z) +F(c, z) = 2+2+2 =6. The **flow** of the network is **6**.

8.4 MAXIMAL FLOW

DEFINITION 8.4.1

A flow F in a network (G, C) is called maximal flow if $|F| \geq |F^1|$ for every flow F^1 in (G, C). In otherwords, a flow F is maximal if there is no flow F^* such that$|F| < |F^*|$.

Example 8.4.2:

The flow in Figure (below) is maximal.

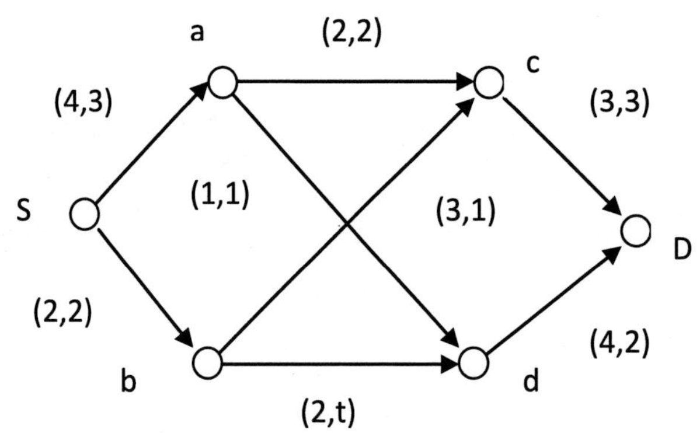

Figure 8.2

The value of any flow defined on the network is such that |F|= F(S, a) +F(S, b)

Since F(S, b) ≤ C(S, b) ≤ 2 and F(S, a) = F(a, d) + F(a, c) ≤ C(a, d) +C(a, c) = 1+2 =3, |F| = F(S, a) + F(S, b) ≤ 5. Then it follows that the indicated flow is a maximal flow.

8.4.3 IMPROVE THE VALUE OF FLOW IN A NETWORK

To improve the value of flow F in a network we start with some initial flow and iteratively increase the value of flow until no improvement is possible in the given network.

We can take the initial flow to be the flow, when the flow in each edge is zero. To increase the value of a given flow, we have to determine a path from the source S to z and increase the flow along this path. Let F bea path from S to z in the underlying graph of the network G. If an edge e in P is directed from node v_i to node v_{i+1}, we say e is properly oriented, otherwise e is improperly oriented.

If we can find a path P from S to z in which every edge in P is properly oriented and the flow in each edge is less than the capacity of the edge, then it is possible to increase the value of the flow. It is also possible to increase the flow in certain paths from S to z in which properly and improperly oriented edges are present. Let P be a path from S to z and let V be a vertex in P different from S and z. Since V is an intermediate vertex in the network, two edges e_1 and e_2 are incident on this vertex V. There are now four cases for the orientation of the edges e_1 and e_2. They are as follows:

1. When the flow is incresaed in each edges by, the flow into V will be equal to the flow out of V.
2. When we increase the flow in e_2 by we must decrease the flow in e_1 by so that the flow into V will still be equal to the flow out of V.
3. When we increase the flow in e_1 by, then we must decrease the flow in e_2 by to achieve equal flow in and out of V.
4. When flow is decreased in both edges by we achieve equal flow in and out of V. All the above cases are possible since $F_{ij} < C_{ij}$ for a properly oriented edge (i, j) and $0 < F_{ij}$ for an improperly oriented edge (i, j). Let $\Delta = \min(x)$, where x consists of the numbers $C_{ij} - F_{ij}$ for properly oriented edges (i, j) in Pand F_{ij} for improperly oriented edge (i, j) in P.

$$\text{Define } F^*_{ij} = \begin{cases} F_{ij} & \text{if (i,j) is not in P.} \\ F_{ij} + \Delta & \text{if (i,j) is properly oriented in P.} \\ F_{ij} - \Delta & \text{if (i,j) is improperly oriented in P.} \end{cases}$$

Then F^* is a flow whose value is greater by than the value of F (i.e., $F^* = F + .$). In every case, the resulting edge assignments give a flow F^*.

8.4.4 LABELLING PROCEDURE

To establish that a given flow F is maximal, we must verify that there are no flow-augmenting paths in a transport network (G, C). For graphs with few vertices and edges we usually determine by inspection that no flow augmenting paths exist. But for larger graphs, inspection alone is not sufficient. For this reason, we introduce a labelling algorithm .

Suppose that we have a network Gin which s is the source, t is the sink and capacity C. The vertices of G are ordered $s = v_0, v_{1,\ldots\ldots,} v_n = t$. In the labeling algorithm, wefirst label the source S and then we label other vertices recursively according to the following scheme.

1. Set $F_{ij} = o$ for each edge (i, j)
2. Label vertex a by (-, ∝) (labeling the source).
3. If the sink t is labeled, got to step 6.

4. Choose the not yet examined, labelled vertex V_1 with smallest index i. Stop if there is no such vertex; otherwise set $v = v_i$. (Go to the adjacent vertices of the source).

5. Let ($\propto$, Δ) be the label of v. Examine each edge of the form (v, w) where w is un labelled. If $F_{vw} < C_{vw}$, label vertex w by (V, min(Δ, $C_{vw} - F_{vw}$)). If $F_{vw} = C_{vw}$, do not label w. For an edgeofthe form (w, v) if $F_{vw} > 0$, label vertex w by (v, min ((Δ, F_{vw}))). If $F_{vw} = 0$, do not label w goto step 3.

6. Let (γ, Δ) be the label of z. Let $w_0 = z$, $w_1 = \gamma$. If the label of $w_i = (\gamma', \Delta')$. Put $w_i + 1 = \gamma$. Continue $w_k = a$. Now the path P: $a = w_k w_{k-1}, \ldots, w_1, w_0 = t$ is the required path from s to t. To change the flow of the edges in P find, if the edge e is properly oriented, then increase the flow in e by Δ. Otherwise decrease the flow in e by Δ. Remove all labels from vertices and goto step 2.

Example 8.4.5: Find the maximal flow in the following network ?

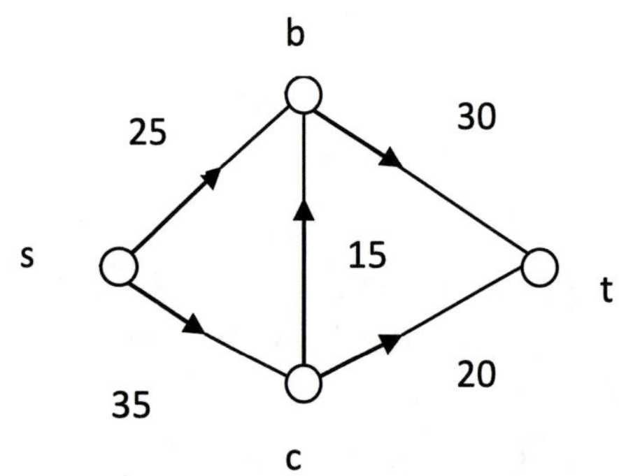

Figure 8.3

Solution

Aranging the vertices in an order we have, as, $V_0 = s$; $V_1 = b$; $V_2 = c$; & $V_3 = t$. Let us Initailly set the flow inevery edge as zero.

1st Iteration: Label the first vertex s as (-, $\propto$). Examine the edges (s, b) and (s, c). For the edge (s, b), $F_{sb} = 0 < 25 = C_{sb}$. Label the vertex b as (s, 25). For the edge (s, c), $F_{sc} = 0 < 35 = C_{sc}$. Label the vertex c as (s, 35). Now the next smallest labeled unexamined vertex is b. examine the edge (b, t). For the edge (b, t), $F_{bt} = 0 < 30 = C_{at}$. Label this vertex t as (c, 25). Since we have reached the sink t, we can stop the labeling procedure and the required path is P: abt. Since all the edges in P are properly oriented, we can increase the flow by. = min (c_{ab}- F_{ab}, $c_{bz} - F_{bz}$) = min (25-0, 30-0) = 25. Thus we have obtained the following network.

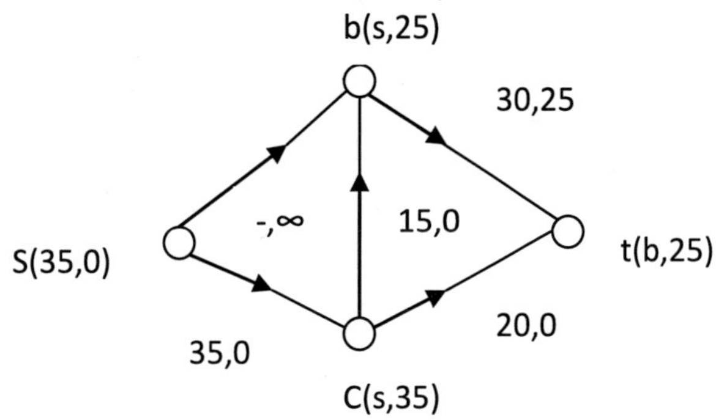

Figure 8.4

2nd iteration: Label the first vertex s as (-, ∝). Examine the edges (s, b) and (s, c). For the edge (s, b), $F_{sb} = 0 < 25 = C_{sb}$. We cannot label the vertex b at this moment. For the edge (s, c), $F_{sc} = 0 < 35 = C_{sc}$. Label the vertex c as (s, 35). Now the next smallest labeled unexamined vertex is e. examine the edges (c, b) and (c, t). For the edge (c, b), $F_{cb} = 0 < 15 = C_{cb}$. Label this vertex b as (c, 15). For the edge (c, t), $F_{ct} = 0 < 20 = c_{ct}$. Label the vertex z as (c, 20). Since we have reached the sink t, we can stop the labeling procedure and the required path is P: sct. Since all the edges in P are properly oriented, we can increase the flow by = min (c_{ac}-$F_{ac,}$ $c_{cz} - F_{cz}$) = min (35–0, 20–0) = 20. Thus we have obtained the follwing network.

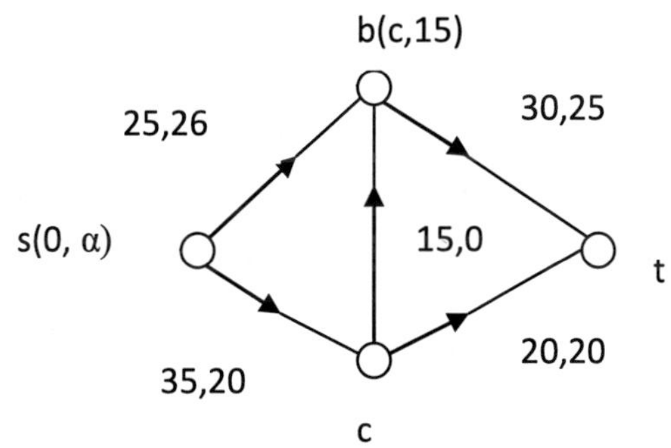

Figure 8.5

3rd Iteration: Label the first vertex s as (-, ∝). Examine the edges (s, b) and (s, c). For the edge (s, b), $F_{sb} = 25 = C_{sb}$. The edge (s, b) is saturated and hence we cannot label the vertex b at this moment. For the edge (s, c), $F_{sc} = 20 < 35 = C_{sc}$. Label the vertex c as (s, 15). Now the next smallest labeled unexamined vertex is c. examine the edges(c, b) and (c, t). For the edge (c, b), $F_{cb} = 0 < 15 = C_{cb}$. Label this vertex b as (c, 15). For the edge (c, t), $F_{ct} = 20 = c_{ct}$. The edge (c, t) is sturated and hence we cannot label the vertex t at this moment. Now next smallest labeled unexamined vertex is b. examine the edge (b, t). For the edge (b, t), $F_{bt} = 25 < 30 = C_{ct}$. The edge (c, t) is saturated and hence we cannot label the vertex t at this moment. Now the smallest labelled unexamined vertex is b. examine the edge (b, t). For the edge (b, t), $F_{bt} = 20 = C_{bt}$ label the vertex t as (b, 15). Since we have reached the sink t, we can stop the labeling procedure and the required path is P: scbt. Since all the edges in P are properly oriented, we can increase the flow by = 5. Thus we have obtained the follwing network.

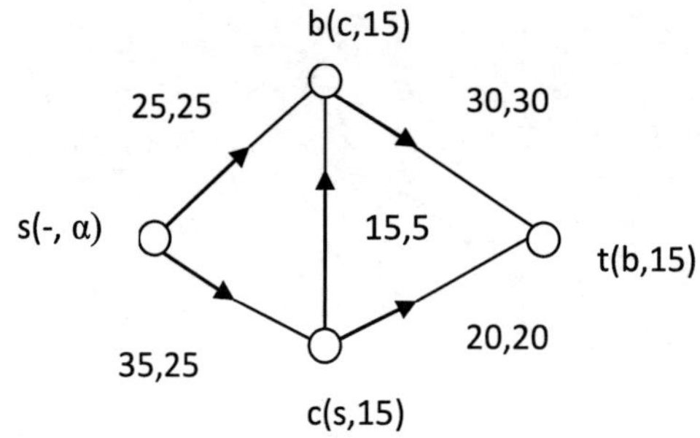

Figure 8.6

4th iteration: Label the first vertex s as (-, ∝). Examine the edges (s, b) and (s, c). Already the edge (s, b) is saturated and hence we cannot label the vertex b at this moment. For the edge (s, c), $F_{sc} = 25 < 35 = C_{sc}$. Label the vertex c as (s, 10). Now the next smallest labeled unexamined vertex is c. examine the edges(c, b) and (c, t). For the edge (c, b), $F_{cb} = 5 < 15 = C_{cb}$. Label this vertex b as (c, 10). For the edge (c, t), $F_{ct} = 20 = c_{ct}$. the edge (c, t) is sturated and hence we cannot label the vertex t at this moment. No further vertices can be labelled. Hence the algorithm ends here. The flow obtained in the previous iteration is maximal.

8.5 MAXIMAL FLOWS AND MINIMAL CUTS

We saw in sections 8.1 and 8.2 that the solution of many practical problem requires finding a maximal flow. Some times, as in example 8.2.2, , a maximal flow can be found quite easily by inspection. Other problems are more difficult. For example it may be that some edges are unreliable and if enough of these fail, then we will not be able to transmit the desired flow. That is, the capacities of theses edges limit the value of the flow. Obviously there are certain limiting values on the possibilies of the value of flow. These suggest the following definition.

DEFINITION 8.5.1

Suppose that (G, c) is a transport network with sources S and sink D. Suppose that X is a set of vertices such that $S \in X$, but $D \notin X$. Let X_1 denote the complement of X in V(G). Then the set (X, X_1) of all edges from a vertex in X to a vertex in X_1 is called an S–D cut.

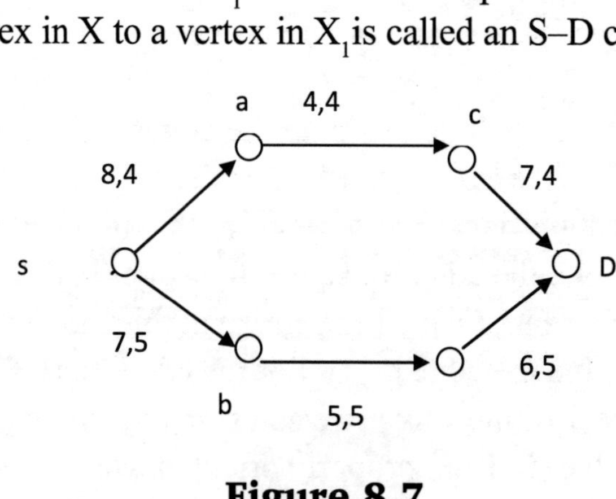

Figure 8.7

In Figure 8.7, the sets X_1= {S}, X_2={S, a, b}, and X_3= {S, a, b, c, d} are examples of S-D cuts. For a transport network with n vertices other than the source and sink there are 2^nS-D cuts (X, X_1).

DEFINITION 8.5.2

If C is any set of edges in a (transport) network (G, C), then the capacity of C is the sum of the capacities of the edges of C. Thus the capacity C(C) is defined by: $Cc(C)=\sum_{e\in C} C(e)$. An S-D cut (X-X_1) is minimal if there is no S-D cut (Y, Y_1) such that C(Y, Y_1) < C(X, X_1) or in otherwords, an S-D cut is minimum, if there exists no S-D cut that has a smaller value.

In general, the value of any flow is limited by the capacities of the cuts. In summary, the value of a flow F can be computed in several apparently different ways.

1. | F | = the total flow out of the source S.
2. | F | = the total flow in to the sink D.
3. | F | = the net flow across any S-D cut.

8.6 THE MAX FLOW-MIN CUT THEOREM

In any Transport network, the value of any maximal flow is equal to the capacity of a minimal cut. This Theorem asserts the following:

1. There exists a minimal cut (X, X_1).
2. There exists a maximal flow F, and
3. For any maximal flow F and minimal cut (X, X_1), | F | = C (X, X_1).

For the network graph in Figure 8.8(a), using the augmented paths (shown in Figure 8.8(b)), s- a- c- d, and s- b- c- d, the value of the flow network is 9 and minimal cut is (X, X_1) where X = { s, b, d}

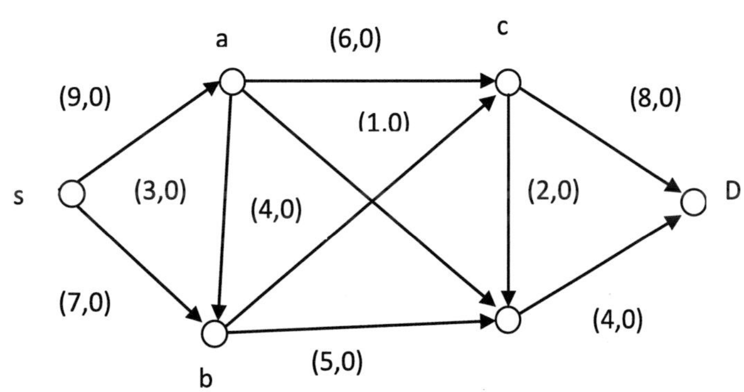

Figure 8.8 (a)

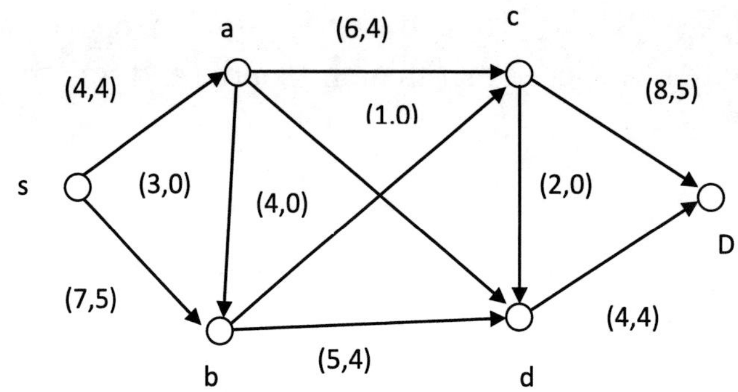

Figure 8.8 (b)

SUMMARY

1. A **flow** network is a directed graph with a distinguished initial vertex (the source) and a distinguished terminal vertex (the sink), in which each edge is assigned a flow capacity. A flow must satisfy the restriction that the amount of flow in to a node equals the amount of flow out of a node, except when it is a source (where outgoing flow is more), or sink (where in coming flow is more).

2. A **Cut** in a Net work is a line that breaks it into two divisions, one containing the Source and the other containing the Sink. The **value (Capacity) of a cut** is defined as the sum of the values on the broken arcs, ignoring the one flowing from sink to Source section. The M**inimum Cut** in a NW is the cut with minimum capacity over all cuts in the NW.

3. The **Maximum flow** of a Network (NW) is the flowwith maximum value, overall flows of the NW. It can be easily found using the **Max-Flow Min-Cut Theorem**, which states that, for any Net Work (G, C), the value of maximal Flow is equal to the Minimum-capacity Cut. That is, the two quantities are equal.

EXCERCISES

1. Find the flow for the following Net work?

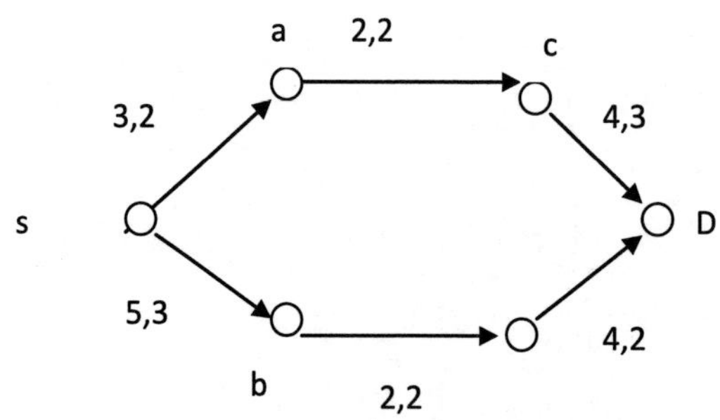

2. Find the Maximal Flow in the following NW by labeling Method?

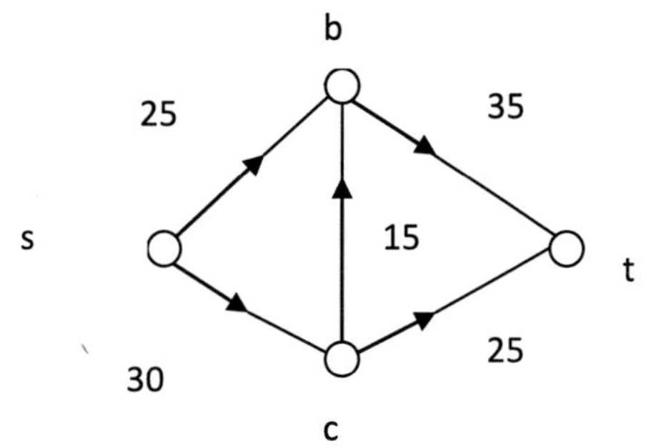

3. For the network shown below with capacities, find (1) two different maximal flows, (2) the value of each of these flows. ?.

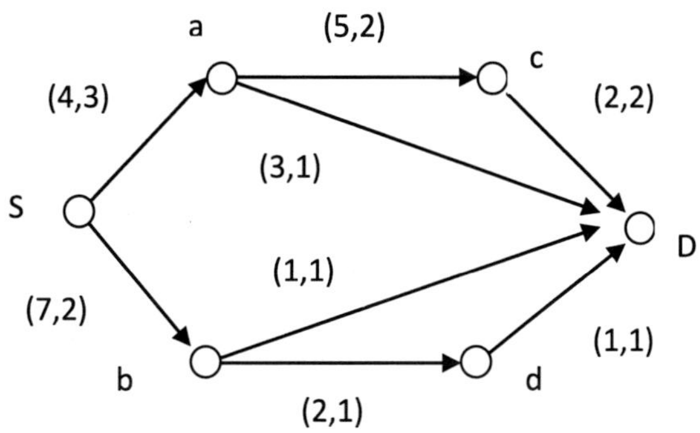

4. In the network with indicated capacities find the maximal flow and all minimal cuts?

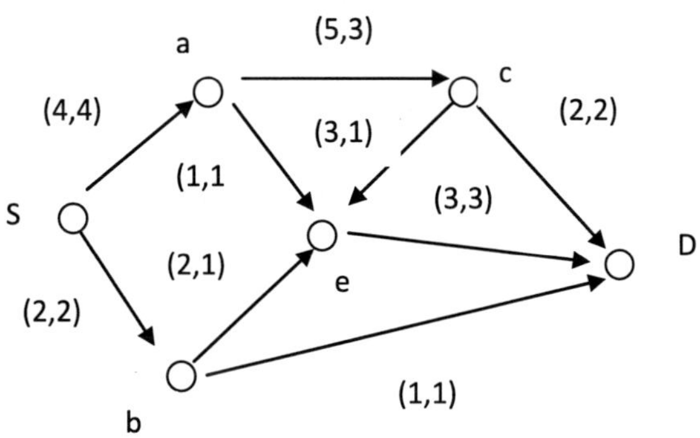

REFERENCES

CHAPTERS 1 AND 2

1. A. C. Aitken., Determinants and Matrices, 9th Edn, Oliver & Boyd Ltd., Edinburgh (1956).
2. C. Berge., The Theory of Graphs And its Applications, John Wiley & Sons, Inc., NY (1962).
3 M. Behzad., G. Chatrand., and L. Lesniak., Foster, Graphs and Digraphs, Prindle, weber & Schmidt Int. Series, Boston, Mass. 1979.
4. C. Berge., and V. Chvatal., Topics on perfect Graphs, Ann. Discrete Maths. 21(1984).
5. C. Berge., Graphs, Elsevier Science Publishers B.V. North Holland, London, 1991.
6. J.A. Bondy., and U.S.R. Murthy., Graph Theory with applications, The MacMillan PressLtd. 1976.
7. G. Chartrand., K.S. Holbert., O.R. Oellermann., H.C. Swart., F- Degrees in Graphs. Ars. Combin. 24 (1987) 133-148.
8. G. Chartrand., P. Erdos., and O.R. Oellermann, How to Define Irregular Graphs, College Math. J. 19(1988) 36-42.
9. J. Clark., and D.A. Holton., A first look at graph Theory, World Scientific Publishing Co . Inc. Teaneck, NJ, 1991.
10. N. Deo., Graph Theory with applications in Engineering and Computer Science, Prentice Hall, 1974.
11. R.P. Gupta., The chromatic index and degree of a Graph, Notices Amer. Math. Soc. 13. (1966), Abstract 66T-429.
12. F. Harary., Graphs And Matrices., SIAM Rev. 9(1)(1967) 83-90.
13. F. Harary., R.H. Norman., and D. Cartwright., An Introduction to the theory of directed graphs., J. wiley & sons, New York. 1965.
14. K.R. Pathasarathy., basic Graph Theory, Tata Mc Graw Hill Publishing Company Limited. New Delhi, 1994.

15. R.C. Read. Graph theory Algorithms, Accademic press, Inc., NY (1970) 51-78

16. F.S. Robertz., Discrete Mathematical Models with application to social, Biological, and Environmental Problems, Prentice Hall, 1976.

17. F.S. Robertz., Graph theory and its applications to problems in society, CBMS NSF Regional Conference Series in Mathematics, SIAM, Philadelphia, 1978.

18. D.P. Sumner., Graphs with I factors, Proc. Amer. Maths. Soc. 42 (1974), 8-12

19. R.J. Wilson., and L.W. Beineke., (eds.), Applications of Graph Theory, Accademic press. 1979.

20. H.P. Yap., Some Topics in Graph Theory, London math. Soc. Lecture Note Series. 108, Cambridge University Press, Cambridge 1986.

CHAPTERS 3 AND 4

1. M. Belmore., And G.L. Nemhauser. , The Travelling Sales Man Problem: A Survey, Operations Res. 16, (1968), 538-558.

2. V. Chvatal., and P Erdos., A note on Hamiltonian Circuits, Discrete Maths. 2(1972), 111-113

3. H. Fleischner., Eulerian Graphs and Related topics, Annals of Discrete maths. 45, North Holland, New York, 1990.

4. S. E. Goodman., and S.T. Hedetniemi., On Hamiltonian walks in Graphs. SIAMJ. Compu. 3 (1974) 214-221.

5. F Harary., and C. St. J. A. Nash Williams. , On Eulerian and Hamiltonian Graphs, Canad. Math. Bulletin.

6. F. Harary., and G.W. Wilcox., Boolean Operations on Graphs, Math. Scand., 20, (1967), 41-51.

7. F. Harary. , and E M Palmer., Graphical Enumeration, Accademic Press, 1973.

8. J.W. Moon., Counting Labelled trees, Canadian Mathematical Congress, 1970 (1965), 701-710.

9. G. J. Minty. , A simple Algorithm for listing all trees in a graph, IEEE. Trans. Circuit Theory, 12(1), (1965), 120.

10. Narasingh Deo., Graph theory with applications to Engineering and Computer Science, Prentice hall, N.J., 1974.

11. E. A. Nordhaus., and J.W. Gaddum., On complementary Graphs, Amer. Math. Monthly 63, (1956), 175-177

12. O. Ore., Note on Hamiltonian Circuits, Amer. Math. Monthly 67 (1960), 55.
13. O. Ore., Graphs and their uses, Random House Inc., NY, 1963
14. S.M. Roberts., and B. Flores., Systematic Generation of Hamiltonian Circuits, Commu. ACM, (1966), 690-694.
15. J.P. Serre., Trees, Springer Verlag, New York, 1980.
16. S. Toida. , Properties of an Euler Graph, J. Franklin Inst. 295 (1973), 343-346.
17. O. Ore., Theory of Graphs, Amer. Math. Soc., Providence, R.I., 1962.
18. O. Ore., A problem Regarding the tracing of Graphs, Rev. Ele. Maths. 6, (1961), 49-53.
19. T. Van AArdenne Eherenfest., and N.G. Delbruijn., Circuits and Trees in Oriented Graphs, Simon. Stevin, 28, (1951), 203-217.

CHAPTERS 5 AND 6

1. A. Battersby., Network Analysis for planning and Scheduling, 2nd ed., St Martins Press, Inc., NY, (1967).
2. H. Bielak., and M.M. Syslo., Perpheral vertices in Graphs. Studia Sci. Math. Hungar. 18 (1983) 269-275
3. F. Buckley., Z. Miller., and P.J. Slater., On Graphs containing a given graph as center. J. Graph Theory 5 (1981) 427-434.
4. G. Chartrand., M., Schultz., and S.J. Winters., On 3eccentric vertices in Graphs. Networks 28 (1996) 181-186.
5. G. Chartrand., W. Gu., M. Schultz and S.J. Winters., Eccentric Graphs, Networks. 34 (1999) 115-121.
6. G. Chartrand., D. Ervin., G.L. Johns., and P. Zhang., Boundary Vertices in Graphs. Discrete Math. 263 (2003) 25-34.
7. G.A. Dirac., Connectivity Theorems in Graphs, Quart. J. Math. Oxford, Ser(2), 3, (1952), 171-174.
8. G. Dirac., Some Theorems on Abstract Graphs, Proc. London Math. Soc. 2 (1952) 69-81.
9. R.L. Graham., and P. Hell., On the History of minimum Spanning Tree Problem. Annals of the History of Computing 7 (1985) 43-57
10. J.P. Georges., and Mauro D.W., On the size of graphs labelled with a condition at distance two. J. Graph. Theory 22 (1996) 47-57

11. F. Harary., The Maximum Connectivity of a Graph. Proc. Nat. Acad. Sci. 48 (1962) 1142-1146.

12. M. Held, and R.M. Karp, The Travelling Sales Man problem and Minimum Spanning Trees, Mathematical Programming, 1, (1971), North Holland Publishing Company, 6-25.

13. P.A. Ostrand., Graphs with specified radius and Diameter. Discrete Math. 4 (1973) 71-75.

14. R.C. Read, Graph theory Algorithms, Accademic press, Inc., Ny, (1970), 51-78.

15. V.K.M. Whittney, Algorithm 422: Minimum spanning Tree, Comm. ACM, 15(4), (1972), 273.

CHAPTERS 7 AND 8

1. A. Battersby, Network Analysis for planning and Scheduling, 2nd ed., St Martins Press, Inc., NY, (1967).

2. E.F. Beckenbach, Network Flow Problems, Applied Combinatorial Maths, John Wiley & sons, Inc, NY, 1964.

3. R.G. Busacker., and T.L Saaty, Finite Graphs and Networks, An introduction with Applications, Mc Graw Hill Book Company, NY, 1965

4. N. Deo., Self Dual Graphs and Digraphs, Proc. Sixth Annual allerton Conf. on circuit and systems theory, Oct 1968, 832-840.

5. A K Dollan., and J Aldous., Networks and Algorithms: An Introductory Approach, Wiley-Interscience, 1993.

6. S Even., Graph Algorithms, Computer Science Press, 1979.

7. L. R. Ford., Jr. and D.R. Fulkerson, Maximal flow through a Net work, Canad. J . Maths. 8 (1956), 399-404.

8. A Gibbons., Algorithmic Graph Theory, Cambridge, 1985

9. L.R. Ford, Jr. and D.R. Fulkerson, Flows in Networks, Princeton University Press, Princeton, 1962

10. T.C. Hu, Network Flows and Integer programming, Addison Wiley pub com Inc, Reading mass., (1969).

11. M. Iri., Network Flow, Accademic Press, Inc, New York, (1969).

12. E.K. Onaga., Optimal Flows in communication Networks, J. Franklin. Inst. 283, (4), (1967), 308-327.

13. L. Lawler., Combinatorial Optimization, Networks and Matroids, 1979

14. T.D. Parson., On Planar Graphs, Am. Math. Monthly, 78(2), (1971), 176-178.
15. M N Swamy., and Thulasidharan., Graphs Networks and Algorithms, Wiley, 1981.
16. W.T. Tutte., Connectivity in Graphs, University of Toronto Press, Toronto, 1966.
17. H. Whittney., Planar Graphs, 21, 1933, 73-84.
18. H. Whittney., Non separable and planar graphs, Trans. Am. Math. Soc, 34, (1932), 339-362.

SUBJECT INDEX

O

P

R

S

T

U

V

W